The BEE

This edition published in paperback in the United States of
America and Canada in 2018 by
Princeton University Press
41 William Street
Princeton, NJ 08540
press.princeton.edu

Library of Congress Control Number: 2018931642

ISBN: 978-0-691-18247-6

This book was conceived, designed, and produced by

Ivy Press
An imprint of The Quarto Group

Creative Director Peter Bridgewater
Publisher Susan Kelly
Editorial Director Tom Kitch
Art Director James Lawrence
Series Commissioning Editor Kate Shanahan
Editors David Price-Goodfellow and Hugh Brazier
Designer Andrew Milne
Picture Researcher Katie Greenwood
Illustrator Sandra Pond

Front cover image: Fotolia/Peter Waters

Printed in China

10 9 8 7 6 5 4 3 2 1

The BEE
A Natural History

NOAH WILSON-RICH WITH
KELLY ALLIN, NORMAN CARRECK
& ANDREA QUIGLEY

PRINCETON UNIVERSITY PRESS
PRINCETON AND OXFORD

Contents

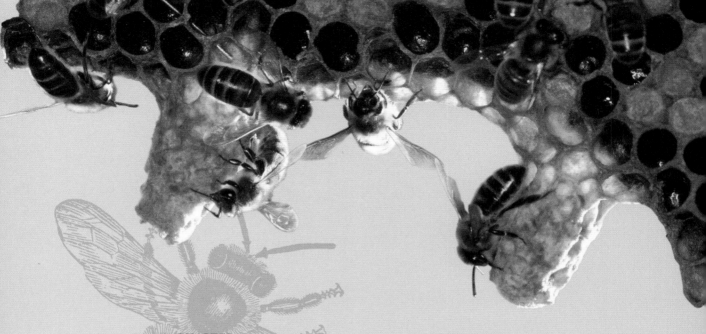

Introducing the Bee

Before flowering plants evolved, there were no bees. And then, about 100 million years ago, plants began to develop colorful appearances and sweetly scented reproductive organs—while at the same time some wasps abandoned their carnivorous hunting lifestyle and took to a gentler way of life. Bees evolved from these wasp ancestors, feeding on pollen provided by the plants for protein in exchange for their services as pollinators.

Bees are remarkable for their co-evolution with flowers, displaying an astonishing range of adaptations—and they are no less remarkable for their social lives. Some bees are solitary, but honey bees live in large, well-organized family groups, and exhibit complex social behaviors seen nowhere else in the animal kingdom. Besides which, honey bees also make several products that are of direct benefit to us—the honey, the wax, and the resins that humans have valued for millennia.

Today, bees are found across the world, and the twenty thousand or so species display an amazing variety of behaviors. Some species live underground, others high in trees, and some even build their nests within the walls of our homes. And of course humans have taken to beekeeping on a massive scale. Bees are now so intertwined with humanity that our interest in them is no longer a simple fascination, but a vital necessity. Quite simply, apart from the honey and other products they supply, we need bees to pollinate the majority of fruit and vegetable crops that we rely upon for our own food.

The challenges faced by bees today—from habitat loss to pesticides and deadly diseases—threaten not only the bees themselves but potentially all of human life.

Right *Bees and other animals cross-pollinate over 90 percent of the world's wild plants.*

THE HARD-WORKING HONEY BEE

There are about twenty thousand bee species in the world and they can be split into four broad groups: solitary, bumble, stingless, and honey bees. While all these groups are covered in this book, the honey bee (*Apis mellifera*) is the most studied and the most familiar.

The honey bee originated in Eurasia and Africa, but is now found on every continent except Antarctica. Honey bees have lived in close association with humans for thousands of years, ever since the Ancient Egyptians established the art of beekeeping. But how many of us, as we pick a jar of honey off the supermarket shelf, pause to consider the lives of the bees that made it?

In a brood cell deep inside the nest, an egg hatches, and the larva, fed first on royal jelly and then on bee bread, grows throughout its first week before developing into a pupa. At this point it is sealed into its cell by the adult workers, and about three weeks after the egg was laid a new worker bee emerges.

The behavior of the adult worker progresses with age. Her first duty is inside the nest, cleaning and nursing the brood, before she moves on to guarding the entrance, and eventually to foraging for pollen and nectar. Foraging takes a severe toll on the worker bee's body, clearly visible in the shredding of her wings. In the summer, a worker honey bee lives only about a month as an adult, but in the inactive winter season she might live for three to six months.

The drones and the queen do not experience such grueling wear and tear, so they have longer lives. Drones can live for three months, or until they finally mate. Honey bee queens are remarkably long-lived. They may survive several years, taking periodic rests during the winter or the rainy season, and then restarting egg laying to produce more workers when flowers come back into bloom.

ABOUT THIS BOOK

This book is for anyone interested in bees, the flowers they pollinate, and the products they create. The first three chapters summarize our scientific understanding of bee evolution, biology, and behavior. We then build on that knowledge to consider how humans have interacted with bees over the course of several millennia, and provide some details of the pursuit of beekeeping, before moving on to an overview of forty of the world's most remarkable bee species. In the final chapter we consider the challenges faced by bees in the modern world, and what we can do to help them. We need bees, and increasingly they need us.

10 FACTS ABOUT BEES

1. Only female bees sting, and many solitary bees can't sting.

2. A bee's sting (or stinger) is a modified egg-laying organ.

3. A bee has five eyes: two are complex eyes that see movement well, while the other three detect light intensity.

4. Bees can see ultraviolet light, but they cannot see the red end of the spectrum, so they perceive the world as more blue and purple than we do.

5. Drones do not have a father, but they do have a grandfather.

6. Bees are herbivores, and their diet comes entirely from flowers—carbohydrates from nectar and protein from pollen.

7. Honey bees are not native to the Americas, and bumble bees are not native to Australia.

8. A queen bee has exactly the same genes as a worker: she develops into a queen simply because she is fed extra rations of royal jelly when she is a larva.

9. The honey bee genome has been sequenced; it is about one-tenth the size of the human one.

10. Bees pollinate over 130 fruit and vegetable crops, and produce many other things that benefit humans—honey, wax, resins, propolis, royal jelly, and even venom.

KEY BEE TERMS

Apiary A place where beehives are kept.

Brood A collective term for the three stages of developing bees (eggs, larvae, and pupae).

Drone A male bee.

Hive The structure (natural or man-made) within which the nest is located.

Larva The developmental stage that hatches from an egg.

Nest A foundation for rearing brood.

Pupa The developmental stage that follows a larva, the period during which metamorphosis into an adult bee occurs.

Pollination Process by which plants undergo fertilization, via pollen transferred from one flower to another of the same species.

Queen A sexually reproductive female (typically only one per nest in social bee species).

Swarm A tightly packed group of honey bees (a queen and many thousands of workers) that flies from the natal nest in search of a new home.

Worker A female bee that is not capable of sexual reproduction, comprising most of the population of the nest in social bee species.

CHAPTER 1

Evolution & Development

The Evolution of Bees 🐝

EVOLUTIONARY PUZZLES

Charles Darwin was fascinated by bees. He considered eusocial ("truly" social) insects such as bees, wasps, ants, and termites to be evolutionary puzzles. The idea that individuals would forgo reproduction to serve a dominant individual (a queen) who is the only reproducing female in the colony was a threat to his theory of evolution by natural selection. Even now, when we understand that evolution relies on individuals passing their genes on to the next generation, it is a puzzle. If organisms do not reproduce, their genes will not be passed on, so this sterile behavior should not persist. But it does. How can individuals that give up reproduction be favored by natural selection? Furthermore, why are these social insects so successful? Some of the answers to these puzzles can be found in Chapter 2.

THE EARLIEST BEES

Over 100 million years ago, the world knew nothing of bees. Dinosaurs walked the earth, while wasps buzzed and flitted about, visiting fern-like plants to prey on smaller insects. In time, however, there was a change in the foraging preferences of some wasps. One particular lineage (the apoid wasps, such as digger wasps, mud dauber wasps, and thread-waisted wasps) began to abandon hunting for animal protein in favor of a vegetarian lifestyle based on flowers. Thus, bees are descended from carnivorous wasps, and over the course of countless generations they evolved to rely entirely on the

Below *Phylogenetic tree showing the taxonomic position of the two most economically important groups of bees: honey bees and bumble bees. After the three domains, only representatives are shown for each group.*

DOMAIN	KINGDOM	PHYLUM	CLASS	ORDER	FAMILIES	GENERA
Bacteria	Protista	Porifera	Crustacea	Coleoptera		Bombus (Bumble bee)
	Animalia				Tenthredinidae	
Eukaryota		Arthropoda	Insecta	Hymenoptera		
	Fungi					Apis (Honey bee)
Archaea	Plantae	Chordata	Arachnida	Lepidoptera	Apidae	

tantalizingly sweet nectar, rich in carbohydrates, and the protein-rich pollen offered by the flowers as a reward to their pollinators. To attract the pollinators, flowering plants in turn developed attractive, scented, and brightly colored flowers, as flowers and bees co-evolved, one inadvertently helping the other.

EVOLVING ALONGSIDE FLOWERS

What makes bees so remarkable is their deeply ingrained relationship with flowers. The spread of flowering plants (angiosperms) across the world coincided with the appearance of the first bees, and together the two very different life forms each facilitated the diversification of the other. Slight changes in the genetic code of a plant species would favor a subset of bees, and vice versa. This co-evolution is a form of adaptive radiation, whereby a single ancestor species spreads out through its progeny into many descendant species, resulting in a blossoming of species diversity—in this case, a wide variety of different species of bees and flowering plants.

Early bees resembled their carnivorous wasp cousins, and had short tongues and sleek bodies. As bees evolved they developed longer tongues that could reach deep down into the hearts of even the deepest flowers to obtain the nectar. Some flowers developed longer flower tubes that only certain species of bees could utilize. Bees' bodies became hairy, enabling the more efficient collection of pollen, and some, including honey bees, developed specialized structures such as pollen baskets and brushes. Some bees today show stunning examples of specialization.

Right *An American southeastern blueberry bee using its long proboscis to obtain nectar and pollen from a blueberry flower.*

SPECIALIZATION

Some bee species are highly specialized in the plants they feed on and pollinate. The American southeastern blueberry bee (*Habropoda laboriosa*), for example, has evolved to primarily feed on nectar from blueberry flowers. The bee is uniquely adapted to home in on blueberry blossom, and its straw-like tongue or proboscis is shaped to suck up blueberry nectar. This bee's real secret, however, is in its buzzing. As it feeds, it vibrates rapidly against the flower's anther to "buzz" pollen off. The pollen sticks to the forager's furry body and is carried back to the bee's nest to provide a food source to developing bees. It also pollinates the next blueberry plant. The co-evolutionary relationship has ensured these bees are primarily active when blueberry blossoms are abundant.

For more about buzz pollination, see page 49.

Honey Hunting & Beekeeping

Early humans relied on hunting wild animals and gathering vegetables and fruits, and in the course of their hunter-gatherer lifestyle they would have come across honey in bees' nests high in the trees. Bees provided our ancestors with perhaps their first condiment—honey. This complex concoction was alluring, and the desire for it drove men to work in groups to capture the golden prize. Their stories are documented in art, and cave drawings from the late Stone Age show a deeply rooted association between humans and bees. Artworks dating back as long ago as thirteen thousand years depict amazing feats, with men scaling seemingly impossibly tall trees, risking falls and stings, to pass the sweet comb down to helpers below. At first, this bravery was the only way of harvesting honey, and it was several thousand years before the refined practice of beekeeping was developed.

ANCIENT EGYPT

The earliest known human-constructed beehive, believed to be some three thousand years old, was discovered in Israel—but the Ancient Egyptians were the first known beekeepers. Evidence from cave drawings in Egypt reveals a long history of beekeeping, dating to at least 2400 BCE and thought to go back as far as 5000 BCE. At first the Egyptians relied on wild bees' nests, but their beekeeping practices later became very advanced. Not only did they construct beehives in the form of woven baskets covered in clay, which stayed permanently in one location, but they also used migratory hives, which floated down the River Nile on rafts, producing unique blends of honey as the bees visited the ever-changing riverside flowers. Egyptians favored the honey bees that were the best honey producers in the local environment, and through a process of artificial selection they gave evolution a helping hand to produce a new subspecies, the Egyptian honey bee (*Apis mellifera lamarckii*).

Some of the beekeepers were likely a lower class of workers, who were forced to work with these aggressive bees and return glorious honey to their superiors—and eventually to the pharaohs and the gods. Official guards accompanied other honey harvesters as they explored deep into the surrounding land, searching low-lying bushes and tall trees for wild hives.

Above *A drawing from ancient Greece of a woman with some honey comb.*

Below *A detail of a bee from a mural in the tomb of Seti I of the Nineteenth Dynasty, from the Valley of the Kings, Luxor, Egypt.*

AMERICA

The people of Mesoamerica (Mexico, Belize, Guatemala, El Salvador, Costa Rica, Honduras, and Nicaragua) kept stingless bees for some two thousand years before Europeans brought the western honey bee to their shores. There are likely at least 250 species of stingless bee in Brazil and at least 4,000 bee species native to North America alone, but the western honey bee is not native to the Americas.

The practice of beekeeping with honey bees was introduced to the New World by Europeans in the seventeenth century as a means of sustainable food production. Some honey bee colonies escaped from their beekeepers' management, and flew off into the woods in this unexplored new land. Native Americans named these feral honey bees "white man's flies," as their arrival announced the advance of the settlers, and gave warning of the inevitable conflict over land that lay ahead.

Right *Beekeepers, from "Venationes, Ferarum, Avium, Piscium" (Of Hunting: Wild Beasts, Birds, Fish), engraved by Jan Collaert (1566–1628).*

The Different Bee Groups

The global abundance of flowering plants from about 100 million years ago opened up a world of opportunity for the vegetarian wasps we now know as bees. Adaptive radiation proliferated across the globe and different types of bees evolved, matching the wide variety of environments, habitats, and flowers.

The twenty thousand species of bees are classified in nine families, which can be divided into three broad groups according to the average length of the bee's tongue, or proboscis. The long-tongued bees sip nectar from deep within the floral chambers. The shorter-tongued bees tend to be more ancient, more closely resembling their carnivorous wasp ancestors.

SHORT-TONGUED BEES

1. Andrenidae Mining bees, found in temperate, arid, warm climates. Absent from Australia. Hairy. Wasp legacy visible in facial structure (eyebrow-like colored patterns, males with facial hair), which might be used for recognition of individuals, but Andrenidae have more advanced eyes. Nest in simple soil burrows, or in hives that require multimodal communication to achieve complex architecture.

2. Colletidae Plasterer bees, masked bees, yellow-faced bees, polyester bees. Typically Australian, with some species in South America and a few in North America and Europe. Unique two-part tongue is used to create remarkable nest burrows lined with smooth polyester secretions. Some transport pollen by swallowing and regurgitating it, rather than by carrying it outside their bodies like most other bees, while other Colletidae carry pollen on leg hairs. Can have large ocelli (simple eyes), helping them to see in dim light.

3. Stenotritidae Smallest family of bees. Australian. Similar to Colletidae, but with one mouthpart instead of Colletidae's distinctive two. Nest in burrows. Large and fuzzy.

Above *Ashy mining bee*

MEDIUM-TONGUED BEES

4–6. Dasypodaidae, Meganomiidae, and Melittidae
Primitive families, the least changed of all bees since
the split from wasps. Found mainly in Africa. Tend to be
picky specialists concerning their food source, sometimes
forgoing pollen for flower oils. The Dasypodaidae and
Meganomiidae families have often been treated as
subfamilies of Melittidae.

7. Halictidae Sweat bees. Worldwide. Often attracted to
sweat. Small. Can be colorful. Sting. Nest underground in
soil or sand. Highly social, with some kleptoparasitic (see
Apidae, right), but there are also solitary Halictids. Their
range of social systems makes them a fascinating group to
use for studying the evolution of social living.

LONG-TONGUED BEES

8. Apidae Cuckoo, carpenter, digger, bumble, stingless,
and honey bees. Worldwide. With or without functional
ovipositor (sting). Solitary to eusocial, some even
parasitizing other bees by stealing their resources
(kleptoparasitism)—such as cuckoo bees, which lay their
eggs on pollen collected by their host bees. Some make
products that are used by humans.

9. Megachilidae Leaf-cutter bees, mason bees,
orchard bees, carder bees. Worldwide. Named for the
materials they gather to make their nests. Carry pollen
in a specialized structure (scopa) on the belly rather than
on hind legs or internally, limiting their ability to carry
pollen efficiently back to the nest and thus requiring more
frequent trips to flowers. Their behavior within the flower
also covers them in pollen. Combined, these factors make
them particularly effective pollinators.

Above *Iridescent green bee*

Above *Bumble bee*

The Evolution & Development of the Honey Bee

Honey bees are living fossils. They comprise the genus *Apis* within the otherwise extinct tribe Apini. *Apis* includes the well-known western honey bee (*A. mellifera*), as well as similar Asian species *A. cerana*, *A. koschevnikovi*, and *A. nigrocincta*. Other eastern species include the small *A. andreniformis* and *A. florea*, and the giant *A. dorsata*. Most honey bees originated in Asia, but recent insights from sequencing the genome of *A. mellifera* indicate that it evolved in Africa, and then spread at least twice into Europe. The honey bees are closely related to other members of the subfamily Apinae, which includes bumble bees, long-horned bees, orchid bees, and digger bees. Those bees have small nests with one to a few hundred individuals, whereas honey bees evolved to live in complex, multi-layered hives, specializing in the long-term storage of honey to feed tens of thousands of individuals.

THE BROOD CYCLE

The eggs of a honey bee look for all the world like grains of white rice. The queen bee lays just one egg in each cell in the brood comb, and after three days the eggs hatch into larvae—tiny worm-like creatures that are as white as snow. Worker bees feed the larvae, which in due course develop into pupae—mummy-like pharaohs that undergo metamorphosis while enclosed in the privacy of a capped cell in the wax comb, until each finally emerges, two weeks later, as a fully grown worker bee.

Fig. 1 *A recently hatched larva at the bottom of a brood cell.*

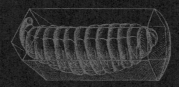

Fig. 2 *The larva grows until it fills the brood cell.*

Fig. 3 *When fully grown the larva turns into a pupa.*

THE QUEEN

Just a few eggs are laid in specially constructed queen cells. These eggs are destined to become queen bees. The queen's DNA is no different from that of her worker-bee sisters; the only difference is environmental. Royal jelly is a secretion produced by the workers and fed to all larvae for a short time, but the queen is fed extra rations, enabling her reproductive organs to develop fully. Simply because of a change in diet, her life is changed forever. She will live for years, rather than a month, and she may lay several hundred thousand eggs over her lifetime, rather than none.

The queen bee is larger than the workers. She emerges from her cell soft and fuzzy and remains that way for a day or so, allowing her body to darken and harden, as all newborn bees must. Within a few days of emerging from her cell, she will take off on the first of up to three mating flights. She will fly high to find her mates in areas where drones congregate. Honey bee queens are notorious for mating with more males than possibly any other animal on earth—genetic analyses have shown up to 29 different paternal lines within one queen's hive. Once mated sufficiently, the queen becomes a slave to her hive, remaining in it for the rest of her life.

Wild Bees Worldwide

Bees are found in nearly all terrestrial habitats worldwide, except in Antarctica and on the most barren mountaintops. They seem to thrive in our own urban environments, as well as in the Arctic Circle, where they take advantage of a seasonal superabundance of flowers. One major distinction across all bees is their relationship to flowers, with some bees specializing on particular plant species while others are generalists. Bees also differ in their habitats, with different foraging and other behavioral traits associated with each habitat type. Another key difference is between tropical and temperate bees.

NEW WORLD BEES

The Western Hemisphere once contained massive and untouched land masses, ripe with diverse yet connected habitats. Left largely unexploited by humans until recently in evolutionary terms, these vast lands were saturated with an unimaginable diversity of flowering plants. The diverse flora of the Americas allowed for thousands of unique bee species to evolve that specialized on native flowering plants.

The tiny squash bees, *Peponapis* and *Xenoglossa*, both of which are members of the long-horned bee family Eucerini, limit their pollen and nectar foraging to squashes, melons, and pumpkins. So intertwined are these bees and flowers that squash bees may occasionally live inside the flowers, timing their activity according to when the squash flowers are open and closed.

The loosestrife bee (*Macropis* sp.) has an affinity toward loosestrife flowers, but because these flowers are not rich in nectar it must be more of a generalist, foraging for other sources of carbohydrates for a balanced diet. Others, such as the striking blue orchard bee (*Osmia lignaria*), have a catholic diet of whatever blooms happen to be in season.

Left *The extended tongue of this female* Peponapis *bee can clearly be seen as she sips nectar from a yellow squash flower.*

OLD WORLD BEES

In Africa and Asia, as elsewhere, persistent threats from predators shaped the evolutionary path of bee behavior, most often by forcing bees to stay on the move rather than settle down and invest in a sedentary lifestyle. Tropical areas tend to host more predators than temperate climates, and these are the habitats where we note more aggressive bees and wasps today. The gentle bees got picked off, or selected out, a long time ago in evolutionary history, posing a selection pressure that favored more aggressive bees in tropical environments. However, as in the Americas, bees were able to diversify alongside the flowering plants, to find safe havens in all sorts of habitats, ranging from sub-Saharan Africa to Siberia and nearly everywhere in between.

The honey bees used for beekeeping by the Ancient Egyptians were notably aggressive, and likely gave rise to a need for humans to perform selective breeding to artificially promote more desirable traits, such as a lack of aggression and low rates of swarming—as noted in the more temperate honey bee races in Italy and Russia.

Below *A fresco from the Tomb of Pasaba in Luxor, Egypt, showing a pillar of the hypaethral court and depicting a scene of beekeeping.*

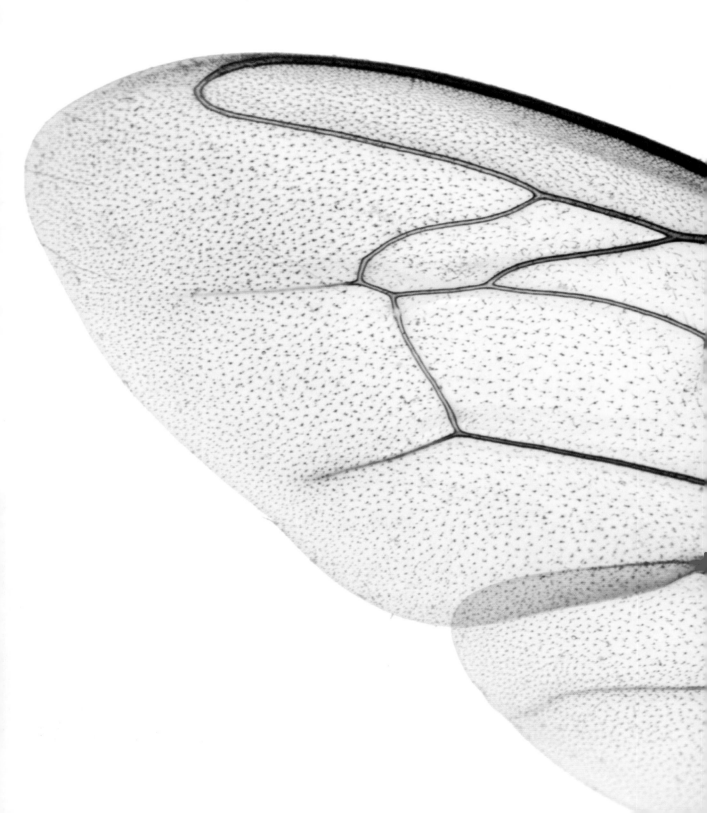

CHAPTER 2

Anatomy & Biology

What Makes Bees Different? 🐝

Bees are unique in many ways. In their anatomy they are similar to their carnivorous wasp ancestors, but in their biology they have evolved into something entirely different. Most bees do not have hardened mandibles (mouthparts) for chewing flesh; they sip nectar from flowers using a specialized proboscis. Bees are not parasitic within other animals like some wasp larvae, but some are social parasites, rather like cuckoos. Bees focus their diet on pollen and nectar, and play a vital role in the pollination of many species of flowering plants. Furthermore, from a human perspective, what really makes bees unique is their significant agricultural, economic, and scientific importance.

AGRICULTURAL

Bees are amazingly effective pollinators, in part because of their sheer numbers. Honey bee colonies have tens of thousands of individuals—perhaps up to eighty thousand—per colony. It only takes one bee to visit, for example, one almond flower, and then a second almond flower, to make an almond. And there are well over a million honey bee hives in the handful of Californian counties that produce almonds for the entire United

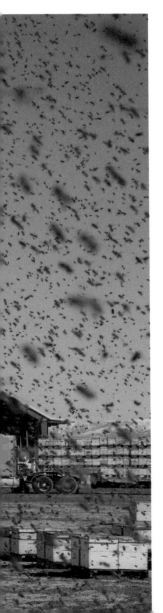

States and regions beyond. Further multiply these numbers by the more than 130 crops that bees pollinate worldwide, and then factor in all the countries around the world growing fruits and vegetables, and you will begin to get a sense of the vital importance of bees to agriculture. These figures also demonstate how massively effective bees are in driving our current agricultural practices. However, it is not just honey bees that are vital to our agriculture; many other types of bee are terrific pollinators too, including bumble bees, mason bees, and squash bees among others.

ECONOMIC

In the USA, honey bees are estimated to contribute over $15 billion annually to the economy. However, the honey bee population has been declining drastically since the 1980s, due to the onset of new diseases and pests, pesticides, and habitat loss, and this decline has coincided with an increase in agricultural demand. The result has been a rise in the price of food, especially in the case of almonds, which up to now have relied entirely on honey bees for pollination. The blue orchard bee (*Osmia lignaria*) has recently been introduced as a pollinator in commercial almond orchards, and other bee species are being studied as possible pollinators for this crop. Bumble bees, too, are used for crop pollination and make a vital contribution to the global economy. In China, a shortage of bees means that human laborers now pollinate some crops by hand. And even in the United States some farmers are turning to human hands equipped with pollination wands and swabs—a technique already used on at least one urban farm in Boston—as a means to guarantee crop yields.

SCIENTIFIC

The research value of bees is enormous, and not only for their contributions in the field of agriculture. Bees can be trained, and the blue orchard bee is a focus of research to train the bees to a target— fruit blossom scent—for increased pollination efficiency. Given that the life span of a worker bee is typically a few weeks to a few months, bees are also used in research relating to age-related disorders such as Alzheimer's disease, studying relationships between aging, memory, and behavior. Bees also act as research subjects in the study of epidemiology, conservation, communication, sociology, genetics, chemical ecology, and more.

Anatomy of
a Honey Bee

Bees are highly adapted to their way of life, and in particular to perform their social functions within the colony. The minute specialization of their functions is also reflected in their anatomy.

To give just one example, worker bees have a barbed sting, while the queen bee has a longer but unbarbed sting.

- The worker's sting is used to repel attackers (and a gland beneath the sting also releases alarm pheromones as the sting is unsheathed, the effect of which is to warn nearby bees of danger); the barb gives the venom gland extra time to pump its contents into the victim—but unfortunately it also leads to the bee's evisceration and death.
- The unbarbed sting of the queen allows her to sting more than once, and she uses it only to kill other queens within the hive and gain her place at the center of the colony.
- Male bees (drones) have no sting—their only social function is to fertilize the queen.

HEAD

ANTENNAE

Set into sockets on either side of the triangular head, the antennae serve as receptors for sensory information and can swivel freely in all directions. They sense smell as well as touch, so can be used to "read" pheromones and flower scents outside, as well as to guide the bee when moving within the darkness of the hive.

THORAX

WINGS

The bee has two pairs of flat wings, the front pair much larger than the hind pair. A row of hamuli, or hooks, on each rear wing lock into a groove on the front wing in flight, holding the two together for strength. Movement is controlled by two sets of muscles that contract in rapid sequence, resulting in strong, fast flight.

LEGS

The bee has three pairs of legs, each leg jointed into six segments. The front legs have brushes used to clean the antennae when they become clogged with pollen. The hind legs carry the corbiculae, or "pollen baskets," that store the pollen.

ABDOMEN

DIGESTION & REPRODUCTION

The abdomen hosts most of the vital organs of the bee: the digestive tract, and, in the drones and the queen, the reproductive system.

STING

The sting is found at the rear of the abdomen in the queen and workers. The worker's sting is barbed and is attached to a poison sac that pumps poison as the bee stings. This organ is a modified ovipositor (egg-laying organ).

PROBOSCIS

Uniquely in insects, the honey bee's sucking tube, or proboscis, is assembled from two other organs, the maxillae and the labium. The bee folds them into the form of a tube when it needs to suck.

EYES

The bee has five eyes—three simple eyes, called ocelli, and a pair of compound eyes, each composed of thousands of ommatidia, or light-sensitive cells, which are used to distinguish light and color and to read direction from the sun's ultraviolet rays.

MANDIBLES

Strong jaws called mandibles that are suspended below the bee's mouth. These can both chew and grasp and are used for chewing pollen, softening wax when building brood and storage cells, and clinging to surfaces.

ARMOR

Like the thorax, the abdomen is protected by armor made of chitin, a cellulose-like substance that forms the exoskeleton of insects.

29

Flight & Internal Anatomy

SECRETIONS

A honey bee has a series of glands that secrete a variety of substances. In the head, mandibular glands and hypopharyngeal glands secrete chemicals used in communication, as well as some food products used to feed the larvae, such as royal jelly. The salivary glands are found in the head and the thorax, and their secretions aid digestion and nutrient uptake. The Arnhart glands on the ends of the legs may be used for marking foraging hotspots and territory. The abdomen contains a rich array of glands. On the ventral side (the underside), wax glands produce wax, and the Dufour's glands produce chemicals that inhibit worker bees from egg laying. The posterior abdomen holds the Nasonov's gland, which secretes recruitment pheromones that draw other bees near, often marking a food source.

BRAIN

The bee possesses a relatively simple three-lobed brain. The brain receives stimuli from sensory organs found throughout the body and processes them. A notable portion of the brain consists of the mushroom bodies—bi-lobed areas that are so-called because they resemble mushroom heads in shape. They are located below and behind the eyes, but above the esophagus, and are hubs for learning, memory, and spatial mapping.

NERVOUS SYSTEM

The subesophageal ganglion controls all motor function of the bee. Interestingly, this is not located in the head, but in the thorax, which explains how the bee can continue walking and breathing without its head. The bee cannot fly without its head, however, because it needs that delicate balancing weight.

CIRCULATORY SYSTEM

Bees, like all insects, have an open vascular system. There are no lungs, and no blood vessels. Oxygen enters the body through spiracles, openings along the side of the abdomen, and reaches the cells directly through a tracheal system that delivers air through the body and returns carbon dioxide to the spiracles. The abdominal pumping movement often noted in bees is merely their breathing mechanism.

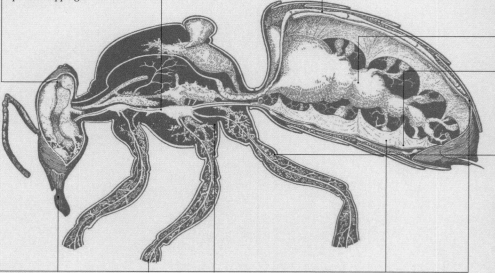

DIGESTION

Bees rely on nectar for carbohydrates and pollen for protein. Nectar is sucked through the tube-like proboscis into the mouth, from where it passes through the esophagus and is collected in the crop—a stomach-like pouch. The honey bee crop carries an average of 20–40 mg of nectar at a time. The proventriculus is a valve that prevents nectar from passing any further, but pollen is able to pass through to the stomach, where nutrients are then absorbed into the hemocoel (open body cavity). Waste is excreted through the anus, with solids passing straight through the digestive tract and dissolved waste from the hemolymph (blood) extracted by means of structures known as Malpighian tubules.

STORAGE

Fat bodies are globular masses throughout the hemocoel that store food, make immune proteins, and may also secrete wax.

MUSCLES

Bee muscles are surprisingly delicate structures, despite the demands of flight, eating, breathing, communicating, metabolizing, and reproducing. Muscles tend to be characterized as either voluntary or involuntary, and are further divided into six categories: somatic, splanchnic, cardiac, dorsal diaphragm, ventral diaphragm, and reproductive.

THE MECHANICS OF BEE FLIGHT

It is often said that scientists have proved that a bumble bee cannot fly, but what was actually calculated, when August Magnan and André Sainte-Laguë studied their flight dynamics in 1934, was that bee wings provide insufficient lift to allow a bee to glide. Clearly, bees do fly, but their mechanism of flight remained a mystery, and apparently a baffling impossibility—the bodies of bees were simply too massive to be supported by their delicate wings. Seventy years later, scientists used high-speed photography to solve the puzzle, only this time with honey bees. They revealed that the shape of the flapping wings produces air vortices at the leading edge, providing the additional lift required.

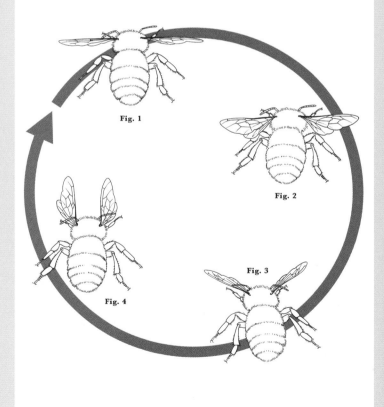

Fig. 1

Fig. 2

Fig. 3

Fig. 4

Bee Senses ✦

While bees share our world, they experience it quite differently. Cues from their surroundings—flowers, the sun, bodies of water—are all received by sensory mechanisms within specialized organs that are very different from ours. So, for example, a flower that appears white to humans may appear blue to a bee.

SIGHT

The eyes of the honey bee have been well studied. They see primarily blues and greens, with some other colors formed through combinations of visual cues. Remarkably, bees see what humans cannot—ultraviolet. Luckily, we can gain insight into what the world looks like to bees by using special ultraviolet filters. Through our eyes, the world looks more yellow and red, while the bee sees it as far more blue and purple. Bees cannot see the red end of the spectrum.

TOUCH

Most bees live inside nests that have little to no light, and where visual communication would be ineffective. Instead, bees rely heavily on vibratory communication. Bees often walk upon one another, exchanging signals by body shaking and, at least in honey bees, by so-called "dancing," as will be explained on pages 60–63. These signals are received through hair receptors throughout the surface of the bee body, including the antennae. Responses to touching and feeling vibrations are followed up quickly, and typically with some degree of specificity with regard to what the message is conveying.

TASTE & SMELL

The antennae receive chemical cues from odors. These odors can be airborne, dissolved in liquid (e.g., nectar), or transmitted directly from another bee through a behavior called antennation. As in humans, taste and smell are closely related, yet a mere 10 types of receptors in honey bees determine taste, while smell is determined by 163 different receptors. In other words, smell is far more important to bees than taste.

With their sense of smell, bees detect nearby floral patches and pick up the approach of competitors and predators. Many bees also signal to others of their species using chemicals called pheromones. Many of these signals are released from glands throughout the body, including Nasonov's gland at the posterior end of

the abdomen. Other chemicals picked up by the antennae from the environment include floral lures for pollination, or even eavesdropped cues that were not intended for the bee.

HEARING

Bees do make sounds, but whether these are functional or merely byproducts is unclear. For example, drones occasionally make a short, loud popping sound when they ejaculate. This likely serves no purpose, but instead is a byproduct of his penis fully inflating and then air rushing out of his body as he explodes and dies.

Other sounds are made by air rushing through the spiracles along the abdomen—such as the quacking, tooting, and piping described in honey bees. A queen occasionally makes piping sounds during her earliest days as an adult. More piping is heard when there are multiple queens present, competing to be the last one standing.

Strangely, reports of bees making sounds in the key of A are increasingly common, ranging from piping in A-flat to buzzing in A during waggle dances. Beyond this, the auditory component of the waggle dance actually has been well studied and described.

Above *The head of a honey bee showing the large compound eyes, antennae and proboscis.*

Right *Marsh marigold (*Caltha palustris*) as viewed in normal daylight (top) and how a bee sees it in ultraviolet light (bottom).*

Genetics 🌿

Bees have a very distinctive genetic system, known as haplodiploidy, which they share only with the other Hymenoptera—ants and wasps. Males have one set of chromosomes, and so are haploid, whereas females have two sets of chromosomes and so are diploid. Any unfertilized egg laid will develop into a male, with half of his mother's genes. In honey bees, the queen pheromone normally inhibits the development of ovaries in the worker or subordinate females. In the absence of queen pheromone, the ovaries of workers will develop, and they may lay eggs, but since they cannot mate, these unfertilized eggs always develop into drones.

HAPLODIPLOIDY & FEMALE BIAS

Evolutionary puzzles arise from the haplodiploid genetic system, including the curious point that males have no fathers, but do have maternal grandfathers. Moreover, some reproductive females mate with multiple males, in a reproductive strategy called polyandry.

Polyandry occurs in many social bees, such as honey bees, some bumble bees, and others. The relatedness of a polyandrous queen's offspring is skewed, with sisters more related to one another than they are to their parents or to their drone brothers.

This is because sisters share half their genes from the queen, and also a proportion more if from the same father. This means that workers are related to each other on average by more than 50 percent, but to their brothers by a maximum of 50 percent, and thus over evolutionary time, because of natural selection, workers favor the production of other workers rather than drones.

These super-sisters probably led the social structure of social bees to favor females over males in a ratio of three to one, in a remarkably strong case of kin selection.

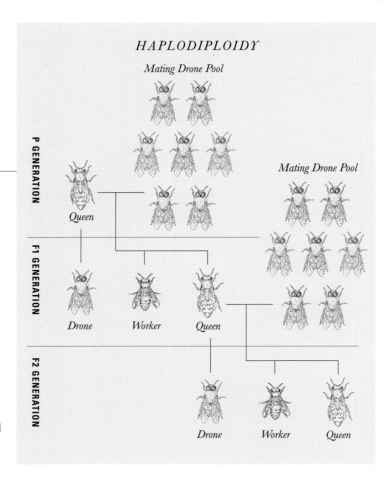

HAPLODIPLOIDY

Mating Drone Pool

Mating Drone Pool

P GENERATION

Queen

F1 GENERATION

Drone   *Worker*   *Queen*

F2 GENERATION

Drone   *Worker*   *Queen*

Above *The haplodiploidy mode of genetic inheritance. Males have one set of chromosomes inherited from their mother. Females have two sets of chromosomes and inherit genes from both their mother and their father.*

GENETIC MISFITS & COMPLEMENTARY SEX DETERMINATION

Within the genetic system of haplodiploidy, sex is determined through the process of complementary sex determination, whereby a single locus in the genome (a specific point on a specific chromosome) determines sex. If there is only one allele (one form of the gene) at this locus, then the individual is a male. If there are two or more different alleles at this locus, then the individual becomes a female. Low genetic diversity, often from inbreeding, results in multiple copies of the same sex-determining allele, and thus genetic misfits. Occasionally, if honey bees are inbred, diploid eggs produce males, and triploid eggs are produced, often with the female phenotype, but on very rare occasions as males, although these are not viable.

There are many examples of even slight genetic abnormalities in the natural world that produce fatal phenotypes. Evidence from paper wasps of the genus *Polistes* shows that viable genetic misfits of these types are produced when genetic diversity is low. If the misfits develop normally, as in some cases with *Polistes*, then the males' sperm is often unreduced when it is passed to the female, and the offspring become triploid. However, unlike its relative *Polistes*, the bee seems to be able to detect genetic misfits at the egg stage, and workers eat them to reuse the protein for a more viable purpose. It is likely that this trait is associated with the bee's advanced eusocial nature, compared to that of their more primitive *Polistes* wasp distant cousins.

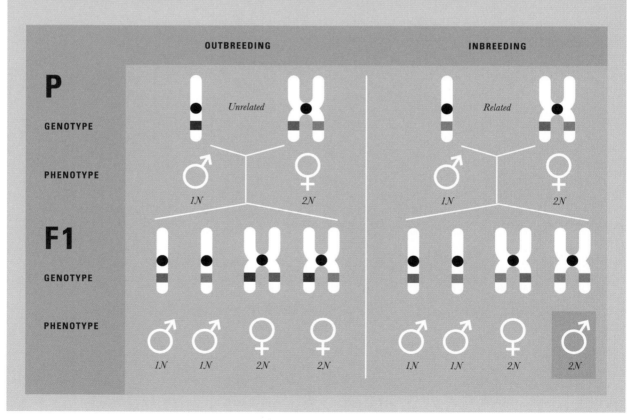

Genomics & Informatics 🐝

Imagine that there was a book for every species, with the chapters providing a blueprint as to how to construct it. Each species' book is its genome. Different editions and printings of the same book can look remarkably different. These can be thought of as individuals of the same species. The chapters and paragraphs of the book arrange how the story is told, as chromosomes arrange how DNA is organized within an organism. Differences at the sentence level resemble alternative alleles at the same locus. Every organism on Earth has a genome, a hidden story to be told, and we are only just beginning to learn how to read it. The human genome was officially completed in 2003. Over 180 genomes of distinct species have been sequenced to date, and the honey bee, published in 2006, was the third insect genome, after the mosquito and the fruit fly.

Below *Honey bee on a DNA fragment analysis map. The honey bee genome is being studied to better understand reproduction, behavior, disease resistance, and more.*

THE HONEY BEE GENOME

The honey bee (*Apis mellifera*) genome sequencing project began work in 2003. It took three years to complete, and cost over $7.5 million US dollars. Its genome contains 236 million base pairs (Mbps), and holds approximately ten thousand genes organized in sixteen chromosomes. Using our book metaphor, this means that the honey bee story is told in 236 million characters, arranged in ten thousand sentences, divided between sixteen chapters.

The completion of the honey bee genome provides an important tool for us to understand how the bee evolved, what makes it similar to other organisms, and what makes it unique. Researchers from all fields of science were drawn to the bee genome as a means to test hypotheses related to the evolution of social living, behavior, and communication.

Prior to the sequencing of the honey bee genome, these bees were thought to have originated in western Asia. Several recent studies using molecular biology have produced somewhat conflicting conclusions, but it now seems likely that honey bees originated in Africa, and spread from there on at least two different occasions, establishing the northern and western European honey bees in one wave, and the southern and central European and western Asian bees separately.

Furthermore, honey bees appear to have evolved slowly; they are so well adapted to their environment that they have changed very little over time. Another remarkable finding in the honey bee genome was that it contains few genes for immune function and detoxification. This means that honey bees rely very little on their own individual immune function (via cells, proteins, and internal chemicals). Given the social nature of honey bees, it may be that most of their resistance to disease comes through their behavior—for example cleaning, grooming, and medicating—rather than from innate immunity.

BUMBLE BEE GENOMICS

The next bee genome is already being sequenced: the buff-tailed bumble bee (*Bombus terrestris*), a bee native to much of Europe and Asia. The estimated genome size is 274 Mbps. This work is taking place at Baylor College of Medicine in Texas, USA. Bumble bees are predicted to have low recombination compared to the extreme recombination in the honey bee. This suggests that bumble bees have produced fewer new alleles from existing ones, and so they remain relatively more conserved and less diverse through evolutionary time compared to honey bees. Furthermore, low recombination may inhibit the bumble bee's ability to adapt to varying environmental and ecological conditions —yet despite the great physical similarity between most bumble bee species they successfully live in a wide range of habitats from the Arctic to North Africa.

Hormones–the
Endocrine System ❧

Bees rely on a system of internal chemical signaling using hormones—chemicals which originate in one location in the body and produce effects at another site. Bee hormones are a hot topic in research today, in part because of their ubiquitous nature, controlling nearly every developmental and behavioral process. How might pesticide exposure inhibit hormone function? What is the tipping point for environmental change to have an irreversible impact on hormones that prevent normal development and function? We may still be years away from addressing these questions fully.

JUVENILE HORMONE

Perhaps the most important hormone in the bee is juvenile hormone (JH). JH was first discovered in the late 1960s, and it is the main hormone regulating and controlling both physiological and behavioral development. It is secreted by the brain's corpora allata directly into the hemolymph (blood). JH levels are highest when the larva is very young, and then decrease as it develops into a pupa. JH is associated with preventing or inhibiting metamorphosis, and therefore any change in its levels can have

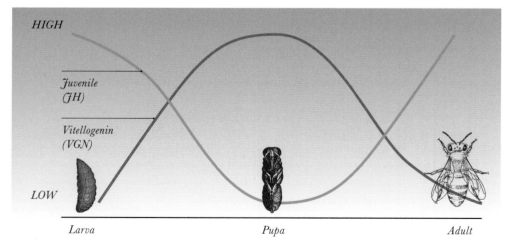

HIGH

Juvenile (JH)

Vitellogenin (VGN)

LOW

Larva   Pupa   Adult

LIFE SPAN OF WORKER BEE

Left *Juvenile hormone and vitellogenin perform a balancing act while the bee develops from an egg to an adult.*

significant impacts on the development of a bee.

Fascinatingly, JH levels rise again as an adult bee ages, suggesting another role for JH in adult behavior. In honey bees, adult behavior is determined by age, in a process called temporal polyethism (time-determined multi-behavior). JH plays an important role in the transition of behavior from helping within the nest to guarding outside the nest, leading, perhaps indirectly, to foraging activity. Aggression levels also tend to correlate with JH levels, with higher levels of aggression seen in bees with higher JH levels.

VITELLOGENIN HORMONE

Vitellogenin (VGN) is a hormone that plays an inverse role to JH. VGN levels in the blood change as the bee matures from egg to adult. Levels tend to increase as larvae develop, triggering pupation. After the bee emerges as an adult, VGN levels then decrease with age. The relationship between VGN and JH is a fine balancing act, with both hormones affecting one another in the form of a feedback loop to control the bee's development. If this relationship is thrown off, then developmental errors could occur.

VGN is found in a variety of animals other than bees that are oviparous (egg-laying), but many functions of VGN appear to be unique to bees. Bees store VGN in their fat bodies, and use it to clear harmful molecules. This especially helps to prolong the life span of the queen, by preventing damage to her body. Queens can live for years, whereas the workers only live for weeks.

In worker honey bees, low levels of VGN facilitate the complex division of labor and the swarming readiness of the hive. Swarming events, like many other VGN-controlled processes, are also related to JH. When JH levels drop, VGN levels rise. Together, these important bee hormones have a direct effect on bee health and life span, both for the individual bee and for the colony as a whole.

Immunology

The immune system of bees is relatively simple. Combined with the diversity of bees, this simplicity allows scientists to test many hypotheses concerning disease resistance and immunology.

How, for instance, might the immune function of a solitary bee compare with that of one living in a complex society? Social bees rely heavily on social interaction, such as grooming and cleaning, while solitary bees rely on their own physiology to defend against infection. Insight from the honey bee genome supports this hypothesis—honey bees have fewer genes for immunology than do solitary species.

Many questions about bee health and welfare have yet to be answered. These are exciting directions for future bee research, with broad implications for science more generally.

INNATE VERSUS ADAPTIVE

All invertebrates rely solely on an innate immune system. It is fast-acting, but it has no memory of prior infection, and it is non-specific in target. Bees do not produce antibodies, nor do they have any so-called adaptive immune system like that employed by vertebrates; their immune system does not learn. The innate immune system relies primarily on three types of barriers: physical, cellular, and humoral.

One aspect of bee immunity remains controversial—the idea of acquired immunity as a type of intermediate form of disease resistance that takes some time to develop, has a degree of specificity against fungal and bacterial infections, and may have some memory. The bee immune system is essentially passive, as far we know. But might acquired immunity be an evolutionary link between innate and adaptive immunity?

PHYSICAL BARRIERS

The first line of defense is behavioral avoidance. But if a bee does come into contact with a potential pathogen, then physical barriers protect it. Bees carry a strong armor, their exoskeleton, made of chitin, which is impervious to nearly everything (except specialized enzymes called chitinases). The bee must have some openings in her body, however, and at these spots we see other, sometimes bizarre, forms of physical barrier. The mouth and anus remain closed when not in use, as do the spiracles. Valves and muscles control all three types of openings, to prevent entry by any potentially harmful substances.

Above Bumble bees are an excellent type of bee for the study of immunology because their large size provides ample hemolymph for testing.

CELLULAR IMMUNE RESPONSE

Should a pathogen or foreign body enter the bee, the second line of defense relies upon cells. The cells that make up the membranes lining the digestive and respiratory tracts provide another physical barrier to infection, but they also go a step further.

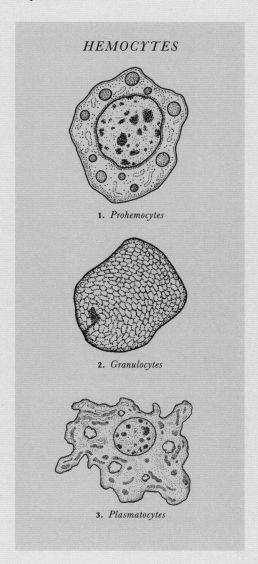

HEMOCYTES

1. *Prohemocytes*

2. *Granulocytes*

3. *Plasmatocytes*

Hemocytes (bee blood cells) come in at least three types:

1. **Prohemocytes** are stem cells that can differentiate into other types of cells, as needed.
2. **Granulocytes** contain granuoles that are dumped at the site of a foreign body or microbial invader. These granuoles can make the hemolymph viscous and inhospitable to the invader.
3. **Plasmatocytes** are able to phagocytose (eat) small pathogens, such as bacteria and fungi, or they can work en masse to surround a larger foreign body, flattening out and hardening to become lamellocytes, which encapsulate and isolate the invader.

HUMORAL IMMUNITY

Should the first two lines of defense fail, the bee relies on a barrage of proteins and chemicals to defend itself against infection and disease.

Phenoloxidase is perhaps the most important enzyme for this function. It converts food (the amino acid, tyrosine) into other products, including toxins (quinones, hydrogen peroxide, reactive oxygen species) that can kill pathogens indiscriminately. Additionally, peptides (short chains of amino acids) with antimicrobial properties are produced in response to particular markers on the cell walls of fungi and bacteria. There are also particular classes of proteins that can kill with some specificity.

Once they have been killed, pathogen fragments and foreign particles are cleared through the Malpighian tubules, kidney-like filtration devices that help cleanse the hemolymph and excrete waste through the rectum.

Pests & Diseases ❧

A myriad of pests and diseases cause infections in bees, which are under constant threat from fungi, bacteria, viruses, protozoa, and arthropods. Some of these affect just one bee species, while others may have jumped from one bee host species to another, making this field of research both fascinating and also challenging.

FUNGI

Bees commonly pick up fungal infections. This is hardly surprising, given their association with flowering plants and the time they spend on damp soil and in moist crevices. Fungi of the genus *Nosema* are widespread, infecting several different bees and causing reduced food stores and colony production. This fungus is also associated with spring dwindle, dysentery, and colony collapse disorder in honey bees.

Fungal spores can get into the bee through openings such as the mouth or the spiracles, or they may land on the bee exoskeleton and germinate there.

In damp conditions, especially in winter, mildew and mold can also break out at the margins of the nest, spoiling pollen stores and the comb. These are not bee pathogens as such, but they can be harmful in a weak colony.

SELECTED BEE INFECTIONS

BROOD (LARVAE & PUPAE)

- Black queen cell virus
- Sacbrood virus
- Acute bee paralysis virus / Israeli acute paralysis virus / Kashmir bee virus
- Slow bee paralysis virus
- Stone brood (Aspergillus fumigatus, A. flavus, A. niger)
- Chalk brood (Ascosphaera apis)
- European foulbrood (Melissococcus plutonius)
- American foulbrood (Paenibacillus larvae)
- Varroa (Varroa destructor)

ADULT BEES (NURSES THROUGH TO FORAGERS)

- Bee virus X
- Bee virus Y
- Acute bee paralysis virus / Israeli acute paralysis virus / Kashmir bee virus
- Deformed wing virus / Kakugo virus / Varroa destructor virus
- Chronic bee paralysis virus
- Slow bee paralysis virus
- Apis iridescent virus
- Amoeba (Malpighamoeba mellificae)
- Nosema (Nosema apis, N. ceranae)
- Varroa (Varroa destructor)
- Tracheal mite (Acarapis woodi)

Key

■ Virus　　■ Fungus　　■ Bacteria　　■ Protozoan　　■ Arthropod

Fig. 1

Fig. 2

Fig. 3

BACTERIA

Bacterial infections are typically more serious than fungal diseases, with a higher incidence of colony loss. Pathogenic bacteria tend to enter the bee orally, and then establish within the gut. Bacterial infections may lead to diarrhea, foulbrood disease, and other diseases. American foulbrood results in rotten, smelly, and liquefied larvae and pupae. Bacterial infections may be resisted by some colonies of bees, but they then infect other colonies or they may be fatal and highly contagious to nearby bees of the same species. For this reason, the method of clearing foulbrood infections from managed honey bee hives is typically fire.

VIRUSES

Viral diseases of bees are highly diverse and widespread, and an important threat to bees today. Certain viruses are closely associated with colony collapse disorder in honey bees. They may be transmitted directly between bees, or through an arthropod vector, typically a mite.

Some diseased bees are easy to spot, with startling appearances including crumpled wings and paralyzed behavior. Many viruses do not cause any obvious symptoms, but nonetheless they adversely affect the health of the bee and may shorten its life.

ARTHROPODS

Arthropod pests are perhaps the best-known threat to bees. They are important because of the diseases they carry. The *Varroa* mite, for example, weakens the host—but it is the viruses it carries that do the damage. Other pests include other mites, beetles, and moths that thrive on bees and/or their products.

The arthropods tend to be invasive species. For example, the mites *Varroa destructor* and *Tropilaelaps* originated in Southeast Asia, where they affected Asian honey bee species, but then crossed species into the western honey bee and have slowly spread. The tracheal mite (*Acarapis woodi*) is an internal parasite that has been a major pest in the USA, causing much damage since it first arrived from Mexico in the 1980s. The small hive beetle (*Aethina tumida*) from Africa, has had devastating effects on western honey bee colonies.

In cases like these, the original host species has usually developed some form of defense or immunity. But options for controlling the new challenges in western honey bees are often limited.

Reproduction 🌿

The reproductive organs of female bees are similar in both solitary and social species. The ovaries hold a lifetime's supply of egg cells that are derived from ovarian tissue, in minute, undeveloped form. As eggs mature, they leave the ovaries and travel down ovarioles, convoluted sets of tubes, where the eggs grow larger before eventually passing out through the ovipositor (the egg-laying organ).

Sexual organs are highly integrated with the bee's digestive tract. When a worker honey bee stings, her barbed sting remains in the victim, leading to her death by evisceration. When a drone ejaculates, his entire penis (or endophallus) explodes into the queen, with such force that his vital internal organs explode with it, leading to his death.

MATING

Courtship behavior is diverse among the bees. Some species indulge in intricate displays to woo a mate, ranging from the private dances performed by large and lumbering male carpenter bees and tawny mining bees, to the public displays of male-on-male competitions seen in honey bee drone congregation areas, high above landmarks such as prominent boulders or church steeples.

After mating, most females are able to store sperm for their lifetime in a special internal organ called a spermatheca. Pre-copulatory choice is limited in honey bee queens, with the fastest drones securing mating opportunities. It is possible that

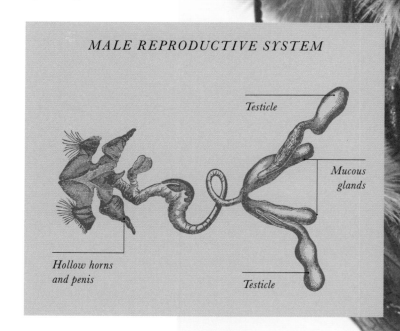

MALE REPRODUCTIVE SYSTEM

Testicle

Mucous glands

Hollow horns and penis

Testicle

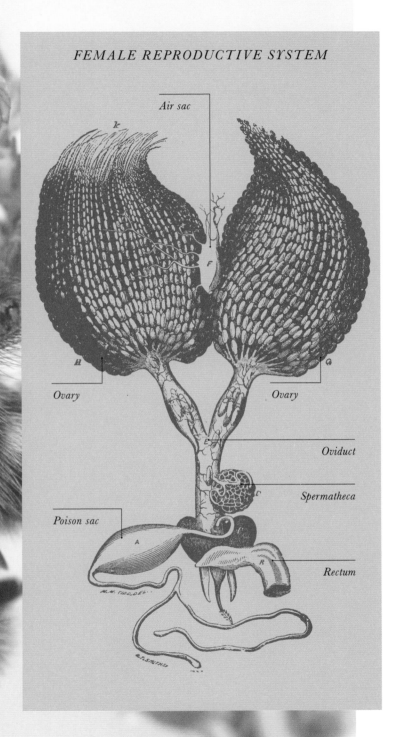

FEMALE REPRODUCTIVE SYSTEM

Air sac

Ovary

Ovary

Oviduct

Spermatheca

Poison sac

Rectum

a female honey bee is able to engage in post-copulatory choice, however. If she has mated with multiple males, it is thought that she can select a particular mate's sperm to fertilize her eggs—but this is controversial, and remains a hot topic of research.

WHY GIVE UP REPRODUCTION?

W. D. Hamilton's work in the 1960s reconciled Darwinian evolution and the worker bee paradox. Hamilton's hypothesis was based on an equation, often referred to as Hamilton's inequality, which looks like this:

$$Br > C$$

This means that helping behavior will be favored when the benefit (B) to related individuals (r is a measure of the degree of relatedness) is greater than the costs to an individual's self (C).

Workers forgo their own reproduction because by helping the queen they can get their genes into the next generation more effectively than if they themselves reproduced. The details of how this works lie in the degree of genetic relatedness among the bees in the colony, which results from haplodiploidy (see pages 34–35).

The Bee's Life Cycle

Bees undergo complete metamorphosis (holometabolism). Unlike some other insects, which repeatedly shed the exoskeleton in a series of molts as the larva grows in size through successive developmental stages (instars), bees expand in size over time within the larval stage, and then enter a pupal stage within a transformative cocoon. They emerge from the pupal casing as a fully grown adult. The timing and other details of development differ among bee families, and within a species they depend on the social role (caste) of the bee.

COLONY LIFE CYCLE

Environmental factors determine when a nest can be founded, when brood is reared, and when the reproductive males are allowed to persist. Tropical bees tend to work around the rainy and dry seasons, while many temperate bees live and die by winter's grasp. Honey bees, however, sit out the winter by clustering within the hive and combining their body heat.

A reproductive female is required for any bee species to found a new nest. She arrives alone for most bee species, or with a swarm of her relatives for species that live in large colonies. The reliance of bees on flowers for food plays a big part in determining the life cycle of the colony. Flowers tend to burst with nectar and pollen after periods of heavy rain, and bees follow suit, founding new colonies just after the heaviest of rains.

Below *The various stages in the life cycle of a bee, from egg to adult.*

Eggs   *Egg showing micropyle*   *Micropyle*   *Larva*   *Pupa*   *Adult bee (Drone)*

Above *A honey bee (Apis mellifera) worker emerges from a honey comb frame.*

In parts of the world where winters are cold, flowers and bees tend to emerge together with the arrival of spring, headed by reproductive females emerging from their overwintering sites.

Bees tend to nest near one another, probably because this increases access to genetically diverse mates. The colonies will grow in size as long as floral resources can support them, or until after reproductive bees are produced, and then disperse to form new nests.

INDIVIDUAL LIFE CYCLE

The life cycles of bees vary tremendously, depending upon species, sex, caste, and environmental factors such as temperature, food availability, predation risk, and

disease. In general, the queen bee, or solitary female, will create a nest, either alone or with helpers, that contains ample resources to support her brood.

Honey bee eggs hatch into larvae that are fed on brood food secreted by the workers, which contains protein and carbohydrates from pollen and nectar. Solitary digger and mining bees lay their eggs on top of food stores that are already provisioned.

Female larvae tend to develop faster than males, although this can vary across species, depending on which sex is larger and needs more time to develop. In temperate habitats, males tend to die off in the fall, and only reproductive females overwinter, to produce a new nest and lay new male brood in the spring.

Pollination

Pollination is the transfer of pollen from male to female flower parts, enabling fertilization. Some plants are self-pollinating, while others require cross-pollination, from one plant to another, through a vector such as wind, water, or an insect.

Many trees and all grasses—including common cereal crops such as rice, wheat, and barley—are wind-pollinated and thus do not need bees. But other plants depend on insects, hummingbirds, and bats for pollination. Many insects perform this service, but throughout the world bees are the most important pollinators.

Above *A honey bee laden with pollen. The pollen is carried in the corbiculae, or "pollen baskets," on the hind legs.*

HOW POLLINATION WORKS

Pollination occurs when male pollen grains from the anther of a flower are transferred to the female stigma. Once the pollen grain lands on the stigma, it forms a tube through which the male gametes travel to reach the ovary. There the male and female gametes fuse, and in due course the fruit, nut, or seed of the plant begins to develop.

Pollination vectors (pollinators) such as bees are crucial to the survival of many plant species, which either rely completely on them or depend on them to assist in the process. Low levels of pollination will lead to reduced yields and fruits that are smaller and slower to mature. If pollinators were to vanish, then most flowering plants and the many organisms they support would disappear.

BEE POLLINATION

Bees do not intentionally provide pollination services. They do it by incidentally brushing pollen grains off an anther and transferring it to a stigma.

Pollen is a protein-rich source of food for bees. When a bee is harvesting pollen, or visiting a flower to drink nectar, the

Left *A bee grips a tomato plant in order to shake the pollen loose.*

pollen grains become stuck to body or leg hairs by electrostatic forces. Bumble bees and honey bees use stiff hairs on their legs to groom most of the pollen into specialized pockets on their legs or body, but leave some pollen coating the body. Many solitary bees do not groom themselves, and simply carry the pollen from plant to plant. It is the pollen left on the bee's body that is rubbed onto the surface of the next flower. Since bees tend to focus on collecting pollen from one flower species at a time, there is a high probability that they will transfer it either to the stigma of the same flower or to that of another flower of the same species.

BUZZ POLLINATION

Buzz pollination, or sonication, is a technique employed by bumble bees, carpenter bees, stingless bees, and several other types of bee to release pollen grains from certain species of plants. These plants have what are known as poricidal anthers, which when vibrated release pollen through small pores; plants and bees may have co-evolved to develop this form of pollen dispersal. The bee must grasp the flower with its legs or mouthparts and then rapidly vibrate its flight muscles to liberate the pollen for harvesting—much like shaking salt out of a salt shaker.

These plants depend on buzz pollination, because there is not enough power in wind or typical bee pollination to extract the pollen from within the anthers. Buzz pollination is a feature of roughly 8 percent of all flowering plants, including many commercial crops such as members of the genera *Solanum* (e.g., eggplants, tomatoes, and potatoes) and *Vaccinium* (e.g., blueberries and cranberries).

Right *A bee covered in pollen. When visiting another flower, some of this pollen will be inadvertently tranferred to the stigma of the other flower, thus facilitating pollination.*

CHAPTER 3

Society & Behavior

Sociality

Bees are remarkable in that they cover the whole spectrum of sociality, displaying a wide range of different types of social behavior and organization. Some species are solitary, some live together in communal groups, others live together in large family groups, and some live in complex societies where individuals are almost completely subservient to the needs of the social group, even giving up their own ability to reproduce in the interests of the hive.

SOLITARY

Despite significant variation in biology between solitary bee species, there are a number of basic similarities. They are all relatively short-lived, and their lifestyle is based on the newly hatched males emerging from the nest before the females. For example, the hairy-footed flower bee (*Anthophora plumipes*), also known as the plume-legged bee, digs a hole in clay or mud as a nesting cavity, within which she lays her eggs in a neat row, with female eggs at the bottom and male eggs at the top. This prevents females from emerging until all the males have done so. The mating season occurs as soon as the females emerge and access males from nearby nests, and the cycle repeats.

Right *The honey bee (*Apis mellifera*) is the ultimate eusocial species. Here, the queen (centre) is attended by the workers.*

Left *The hairy-footed flower bee (*Anthophora plumipes*) is a solitary species. Here, a male emerges from the nesting cavity to await the emergence of the females.*

SUB-SOCIAL

Euglossine (orchid) bees such as *Euplusia surinamensis* advance solitary living one step further, with individual queens working together. These Costa Rican orchid bees take over old nests of others and, by not building a nest from scratch, have an advantage over competitors because they can rear more offspring.

Metallic-green faced bees (*Augochlorella*) can found nests either solitarily or socially, depending on the species and which sex of offspring emerges first. If males emerge first, nests will be solitary, but if one or two female workers emerge first a social nest is the result. Some *Augochlorella* bees also form nests which have several laying queens.

In other species, cooperative breeding systems involve adults providing significant care to young that are not their own genetic offspring.

EUSOCIAL

Eusocial animals are defined as those that live in highly complex groups, with three additional requirements. These groups must have an overlap of generations, a reproductive division of labor (worker castes versus reproductive castes), and cooperative care of the brood. In other words, these true societies have a reproductively dominant individual (typically a queen), and she lays eggs that are reared by individuals other than their mother—for example, by worker bees. Examples of eusocial bees include honey bees, bumble bees, sweat bees, and carpenter bees.

Eusociality is also found in all ants and termites, some wasps, some gall thrips, several kinds of aphids, one species of ambrosia beetle, one species of snapping shrimp, and even two vertebrate species (naked and Damaraland mole rats).

Pathways to Eusociality 🐚

How did social bees evolve? Wasps, the ancestors of bees, often show a primitive form of eusociality in which workers retain their physiological ability to mate and reproduce. In advanced eusocial species, however, a point of no return has been passed. The workers have much-reduced reproductive organs and are no longer capable of mating.

There are several evolutionary routes that bees and other eusocial animals may have passed through to become "truly social." The many possibilities can be divided broadly into two distinct routes: the bees may have arrived at eusociality via the sub-social pathway; or they may have arrived via the semi-social pathway.

Both pathways start with individuals that cooperate in some way, perhaps by sharing or communicating with one another about resources or habitat. Over generations, this cooperation, which started with acts that were simultaneously beneficial to both individuals, becomes more complex and forms into reciprocal altruism, when the benefit to one of the parties is delayed and the favor is returned later. For example, a bee could lead another bee to a flower, with a subsequent return on this investment when the second bee later does the same. This leads to sociality.

Above *Honey bee (*Apis mellifera*) workers cooperate by exchanging food—a behavior known as trophallaxis.*

SUB-SOCIAL PATHWAY

In the sub-social pathway, solitary female bees start their own nests, and each guards the entrance to her nest. This protects against many ecological factors, including predators, competitors, and perhaps even variation in the environment. Over evolutionary time, this nest guarding increases in intensity to involve guarding by offspring that stay behind to protect the nest, rather than disperse to found their own nests.

This cooperative breeding step is a critical point in evolution. When the young stay at home permanently and never breed—and lose the physiological capacity to breed—that leads to a eusocial species, via a cooperatively breeding intermediate step.

Allodapine bees are noted communal breeders, located primarily in sub-Saharan Africa, Madagascar, eastern Asia, and Australia. These bees are remarkable in their social structure. Their lifestyle today is like a snapshot of evolution in action. Some allodapines are communal nesters, some are primitively eusocial (offspring can reproduce but do not), and some are advanced eusocial (offspring cannot mate and reproduce).

Below *In Texas, a blueberry bee (*Osmia ribifloris*) uses an old nest constructed by a mud dauber wasp of the genus* Trypoxylon.

SEMI-SOCIAL PATHWAY

The semi-social pathway has a communal-breeding intermediate step. Here, solitary nests are grouped or aggregated close together, sharing nest building and defense, while all females reproduce. Dominance hierarchies eventually develop, with one female taking an unequal share of reproduction and other females losing reproductive opportunities. An overlap of generations forms, with young females becoming servants to older, dominant females.

This pattern of social organization may also arise when there is limited suitable habitat for nesting sites. The tawny mining bee (*Andrena fulva*), the red mason bee (*Osmia bicornis*), and some metallic sweat bee species (*Lasioglossum* [*Dialictus*]) use existing nesting sites, such as pre-excavated burrows or old nests. If these are unavailable, then we can well imagine how a shortage of such nest sites could have led to the semi-social pathway.

Reproductive Division of Labor

The concept of giving up one's own reproduction to help the group was a challenge to Darwin's theory of natural selection. An appreciation of the mechanisms by which this occurs within individual life spans helps us to understand the process.

OLFACTORY CONTROL

In many eusocial bee species, the queen utilizes queen mandibular pheromone (QMP), a complex multi-component chemical, to maintain her dominance. QMP attracts workers and inhibits their ovary development, making the queen the primary egg-layer. In addition, her eggs have a special queen scent on them, secreted by the Dufour's gland at the end of her abdomen. In the absence of a queen, a worker bee can produce and emit a pheromone that mimics QMP, inhibiting other workers from egg-laying behavior. This worker then gains the ability to lay eggs, albeit unfertilized ones.

DOMINANCE HIERARCHIES

Primitively eusocial bees are those with workers that are able to reproduce, but do not. They forgo reproduction in the presence of a dominant female. For example, worker sweat bees and worker bumble bees are females that are seemingly capable of reproduction.

Left *Facultative helpers: red-tailed bumble bees (*Bombus lapidarius*).*

PLASTICITY

For the worker bees that are capable of reproducing, such as the sweat bees and bumble bees mentioned opposite, the social nature of the species restricts their opportunities—but at the same time it can bring benefits. If the dominant female should die, a worker that is physiologically capable of breeding may be able to inherit the nest and take over the role of dominant female. This is an example of plasticity.

In contrast, advanced eusocial bees have workers that are physiologically incapable of sexual reproduction. This occurs only in two tribes of bees, the stingless bees (Meliponinae) and the honey bees (Apinae), although these two lineages are thought to have evolved sociality independently of one another. Worker females in these lineages have stunted and non-functioning reproductive organs, and they are therefore incapable of mating. This evolutionary step was a point of no return, beyond which bees were no longer facultative helpers, but became obligate workers, bound to a life of servitude to their queen. The plasticity has been sacrificed for the good of the group.

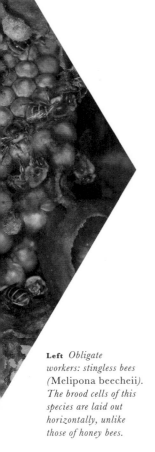

Left *Obligate workers: stingless bees (*Melipona beecheii*). The brood cells of this species are laid out horizontally, unlike those of honey bees.*

WHY DO HELPERS HELP?

The ecological constraints hypothesis is a leading explanation of how non-reproducing helper bees evolved, via the intermediate step of offspring delaying dispersal. If available nest sites are insufficient, not all bees can make nests of their own. According to W. D. Hamilton, this favors the evolution of behavior that helps kin, rather than selfish behavior. Other hypotheses for the evolution of eusociality remain in scientific debate, including group-level and multi-level selection arguments made by E. O. Wilson and D. S. Wilson, the monogamy hypothesis proposed by J. Boomsma, and genetic predisposition ("because their ancestors did it") and pre-condition arguments.

Swarming

Swarming occurs when honey bee colonies divide to start a daughter colony. Half of the original colony stays behind to raise a new queen while the old queen leaves to find a new home with the rest of the colony. The swarm usually remains in one location, perhaps on a tree branch, for up to two or three days while scout bees seek a new nest site—and this is where the swarm will ultimately settle.

Swarming is usually triggered by a lack of queen mandibular pheromone. If this chemical is no longer properly distributed throughout the colony, then worker honey bees will initiate rearing of one or more new queens. This may be due to overcrowding in the hive, or to the age of the queen.

Above If the bee colony is a "superorganism," a swarm is analogous to the colony reproducing, splitting to form two or more colonies.

SWARM INTELLIGENCE

The US biologist Tom Seeley has invested much of his career in studying honey bee swarm intelligence—the solving of cognitive problems through the pooling of knowledge held by the individuals in a group. Each individual in a group has limited knowledge, but the group as a whole successfully makes beneficial decisions. Seeley studied this concept within the context of how honey bee swarms choose a new nesting site. Many of Seeley's swarming experiments took place on an island off the coast of Maine. He set up empty boxes of varying sizes as potential nesting sites for honey bee swarms to move in to. Seeley videotaped the behavior of swarms at an initial site, and his team members recorded the presence and behavior of scout bees at potential nest sites. He hypothesized that the colony makes a decision using a biological process known as quorum sensing. He determined that a quorum of only twenty to thirty bees is required to choose a new nesting site, not a consensus of the whole swarm. This small portion of the colony is optimal because it is large enough for the group to make an accurate decision, but small enough to conserve time, resources, and energy.

SUPERORGANISM THEORY

A "superorganism" is a single functional organism that is made up of many smaller organisms. Several scientists have embellished upon this idea, notably the sociobiologists E. O. Wilson and Burt Hölldobler (2008). This concept is analogous to cells working together to form a single body, with worker bees as the cells, the queen as the reproductive organs, and the scout and guard bees as the skin. Insect colonies, especially bee colonies, may be considered superorganisms. In the case of honey bee swarming, this can be likened to a superorganism giving birth.

TROPICAL VS. TEMPERATE BEES

Honey bees are the only bees that swarm. A "reproductive swarm" often occurs when food is plentiful—for example, springtime in temperate environments. This allows for a full season of gathering food, so that the colony builds up to a size large enough to survive the next winter.

Swarming to produce a daughter colony is seen in both temperate and tropical honey bees, but tropical honey bees may also move en masse or abscond at any time of year, not for reproductive purposes but because food is scarce, or to avoid predators such as hornets. This can occur repeatedly within a few days to avoid predators, and the search for food may cause a colony to migrate several times a year.

The temperate western honey bee will abscond in certain circumstances, but these movements tend to be triggered by exposure to threats such as repeated nest disturbance or fire. When swarming, honey bees are generally not aggressive, as they have no brood to protect.

Right *The swarm waits in a tree while scouts seek out a suitable new nest site.*

Communication ❧

Communication among eusocial species such as honey bees has been extensively studied, but little is known about the vast majority of other bee species.

Honey bees communicate information about distance, direction, and quality of food sources by dancing—the famous round dance and waggle dance, described in more detail on pages 62 and 63.

Sometimes, the messages transmitted by bees are intercepted by other bee species, or by competing colonies of the same species. For example, some species of stingless bees eavesdrop on other colonies in this way, receiving their pheromone signals and then exploiting their food sources.

WAGGLE DANCE RESEARCH—PAST, PRESENT & FUTURE

In 1946, Karl von Frisch, an Austrian ethologist (animal behavior scientist), was studying the color vision of honey bees. He noticed that one bee would arrive at a food source, and then, later, many bees would arrive. He became interested in finding out how the bees communicated the location of the food source to foragers, and discovered that honey bees use dances. Von Frisch observed several distinct dances, such as the "round dance" to communicate the presence of nearby food sources, and the "waggle dance" to communicate the location of more distant food sources. In 1973, Karl von Frisch won a Nobel Prize in Physiology or Medicine for his discovery.

Today, research continues into understanding the details of the waggle dance, and what social factors might influence dance variations and information content. Professor Heather Mattila of Wellesley College in Massachusetts is one such researcher—a "bee whisperer," who can watch a bee's waggle dance and interpret it so that she can go to the floral patch herself. Mattila has found that more genetically diverse colonies are more likely to dance for longer, and that they send more foragers out to find food. This could help explain why honey bee queens mate with so many drones: it is a strategy to increase the genetic diversity of the colony.

The honey bee's waggle dance has not yet yielded all its secrets, however. For example, scientists have yet to figure out how bees effectively communicate using the waggle dance in the darkness of their hives. Evidence from Adrian Wenner's work in California showed that dancing honey bees produce a distinctive sound. Wenner also studied the influence of olfactory cues from the environment, and signals from the Nasonov's gland at the tip of the abdomen. Collectively, these lend support to his theory that dancing honey bees can communicate in the dark by using sounds and smells.

Dance Communication 🐝

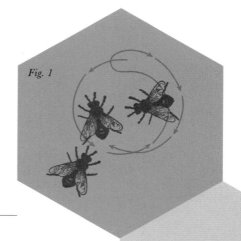

Fig. 1

One of the most remarkable aspects of honey bee behavior is their ability to communicate through dancing. In their dances, the bees transmit information by means of vibrations, sounds, and scents.

Fig. 1 *To communicate the presence of a nearby food source, the bee performs the round dance, walking repeatedly in circles and frequently switching direction.*

ROUND DANCE

When Karl von Frisch began to explore why so many bees eventually arrived at a food source, he first observed the round dance. The round dance consists of the bee walking in circles repeatedly and switching directions. This dance communicates the presence of a food source that is close to the hive (less than 165 feet/50 m away). The dancer repeats the round dance several times, either in the same location within the hive or in different areas. At the end of the dance, she distributes harvested pollen or nectar to the workers following her, so they can taste it and smell out the source—and thus she recruits other foragers to visit the source she has found.

WAGGLE DANCE

The waggle dance is used to communicate the location of a food source that is more than 165 feet (50 m) from the hive. The dance consists of a waggle phase, in which the bee buzzes as she walks in a straight line and vigorously shakes her abdomen horizontally (left to right), and a return phase, in which she walks in a loop back to the starting point to repeat the waggle phase. On the next return phase she turns in the opposite direction, to form a figure-eight pattern.

Three critical pieces of information are encoded in this dance: direction, distance, and quality. The angle of the waggle run, relative to a straight line toward the top of the comb, signals the direction a bee should fly relative to the sun's position (and the bee's internal clock allows her to adjust the direction of the dance as the sun moves through the sky). The duration and length of her waggle run signals how far a food source is from the hive. The enthusiasm of her dance signals the quality of the food, with better food stimulating her to perform a more passionate dance.

Above right *Honey bees (Apis mellifera) returning to the hive pass information about the source of nectar and pollen to other workers.*

Fig. 2

Fig. 2 *For a more distant food source, the bee performs the waggle dance. The angle of the central waggle run indicates the direction of the food source, and the duration and length of the dance gives the distance from the hive.*

CUT THE MUSIC!

As with all languages, there is a dialogue. Sometimes, a member of the colony will head-butt the dancer to stop her dance. This head-butting signals that workers should stay away from the location being communicated because a danger has been spotted there.

The waggle dance is also used when bees are in search of a new home. If a forager finds an optimal location, she returns to the hive and communicates the location to her hive mates via a waggle dance. Head-butting is also used in this scenario by other workers to diminish the dancing bee's activity, either when the message has been received clearly or because the colony has already found a more suitable new home.

Olfaction

Bee antennae contain thousands of sensory components, some of which—on the tips of the antennae—are olfactory receptors. Compared to other insects, bees have many smell receptors, indicating the importance of olfaction within bee communities. Their sense of smell allows bees to recognize fellow hive members, communicate socially within the hive, and find food sources. Some species leave odiferous chemicals on the leaves of a new food source to mark a trail for other hive mates. There is evidence of species-specific codes that determine behavioral responses to different odors.

BEEKEEPING NOTE

Alarm pheromone is released when bees confront enemies. This pheromone is released when a bee stings, and its banana-like scent attracts other bees to assist in the attack. This is why a beekeeper should never eat bananas for breakfast! One method to avoid attack is to use a smoker, which prevents bees from chemically communicating by blocking the reception of alarm pheromones, thereby keeping the hive calm.

The overwhelming majority of research in olfaction in bees has focused on honey bees. Olfaction may be especially important in honey bees, given the combination of a complex social organization and a hive that is typically located in a cavity where there is no light beyond the entrance.

REPELLANTS

The majority of bee species live in very dark and confined habitats. Their resources are gathered in one spot, and could thus make an easy target for robbers—either other bees or a range of different animals. Natural selection favored bees that were able to repel robbers. Olfactory mechanisms of repelling provided an adaptive benefit, given the environment around the food stores. The honey bee's alarm pheromone 2-heptanone, which is released by the mandibular glands, acts on invaders as an airborne and volatile paralytic agent. Once temporarily paralyzed, the would-be robber can be dragged out, and the food stores remain safe.

LURES

Aggregation pheromones are used by workers to attract other workers to important areas, such as a food source or nest site. One example from honey bees smells a bit like almond—and that may be no coincidence, because the almond depends on bees for pollination, although

Left *A bee's antennae contain olfactory receptors, among many other sensory organs.*

BOMB-SNIFFING BEES?

Research has shown that honey bees can be trained to detect explosives using their sense of smell. By providing them with a mixture of sugar, water, and the odor of an explosive, honey bees can be trained to associate the scent of the explosive with food. In response to smelling the explosive, a bee will then stick out its proboscis, signaling the recognition of a food source. Honey bees have the potential to detect explosives in a parts-per-trillion concentration. For information on how bees are being trained to detect land mines, see page 95.

this link is purely speculative. Plants, in turn, have evolved ways to draw these pollinators in, using the sugar in their nectar and their fragrances to lure bees back to them multiple times. Is the bee mimicking the almond, or is it the other way round?

DOMINANCE

The queen utilizes an eight-component pheromone to maintain her dominance. This pheromone attracts workers and inhibits their ovary development, making the queen the sole egg-layer.

The antennae of drone honey bees are more complex than those of females, because they are specifically designed to detect the queen pheromone at times of mating.

Navigation 🐝

After locating a new food source, honey bees return to their hives with samples from the new food source and communicate its location to the rest of the colony (see pages 60–63). But how do they know where it is so that they can inform their hive mates, and how do they find their way back to it? More fundamentally still, if workers were unable to navigate their way back home, then the colony would slowly disappear. Navigation skills are essential for colony survival.

Both social and solitary bees are able to navigate using a variety of senses and a variety of environmental cues: smell, the sun, landmarks, and even electromagnetic waves.

SHORT-DISTANCE NAVIGATION

When navigating on short flights, bees rely heavily on their sense of smell to locate new food sources, and on landmarks when returning to the nest. In social bees, such as honey bees, a forager will bring back pollen for other workers to sample, and will communicate its source so that other foragers can leave the hive to sniff out the flower that it came from. Odors thus attract bees to flowers from shorter distances, but cannot summon them from afar. Likewise, bees cannot rely on their sense of smell to navigate home from a long flight.

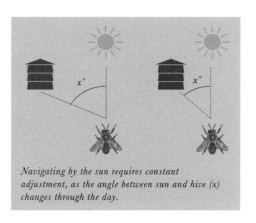

Navigating by the sun requires constant adjustment, as the angle between sun and hive (x) changes through the day.

FAR NAVIGATION

The sun plays a vital role in bee navigation. Bees are able to see the orientation of polarized light, and they only need to see a small patch of blue sky to determine the direction of the sun, since the sun determines the plane of polarization. Every four minutes, the sun shifts one degree to the west. One might think that this would cause bees to get lost, but the consistency of this change, coupled with the bees' circadian rhythm, allows them to calibrate to the constant change in the sun's direction.

Foraging Behavior

A lifestyle completely dependent on flowers is what differentiated bees from their last common ancestors, the meat-eating wasps. Foraging is the term used to describe the behavior of seeking and locating flowers for the purpose of collecting food resources, in the form of nectar and pollen. Bee foraging behavior is fascinating to study, and fun to watch.

OPTIMAL FORAGING THEORY & GAME THEORY

Studies of bee foraging often come down to cost/benefit analyses. How far should the bee travel to get food? How might the presence of other bees change her plans? Scientists use two models to make predictions about foraging behavior.

Below *Hairy-footed flower bee (*Anthophora plumipes*) female feeding on nectar from a flower of meadow cranesbill (*Geranium pratense*).*

THE SCOPA & THE PROBOSCIS

Scopae are specialized structures that carry pollen—the source of protein for most bees. Probosces are similarly specialized tongue-like structures that sip up nectar—the source of carbohydrates for bees. Both of these structures have co-evolved alongside flowers. In the case of specialist bees, the size of the bee's scopa correlates with the pollen size of its preferred flower, and the length of the proboscis depends on how far into a flower blossom the nectar is located.

The first model is optimal foraging theory. It relates the distance traveled to the frequency of encounters with high-quality food items. This produces some fascinating findings, for example that the benefits of low-quality food seem to be negated by the energetic costs and the risk of predation involved in foraging for it.

The second model used to predict foraging behavior is game theory. This factors in other bees, and the costs of encounters with them. If another bee is at a flower patch, then the benefits are decreased by the lower availability of food and the higher risk of conflict.

These tools could be useful in the future, as suitable habitat for bees diminishes and it falls to humans to create forage for them—which will need to be at a practical location for the bees.

PROBOSCIS EXTENSION RESPONSE TEST

The proboscis extension response test is a tool used by scientists to test hypotheses related to behavior and learning. It is a Pavlovian association procedure, whereby bees are exposed to an odor along with a reward. Over a series of trials, the reward is gradually taken away. If the behavior associated with receiving the reward remains— the bee sticking out her proboscis (tongue) to sip sugar water—then the subject has successfully *learned* this association.

Researchers at the United States Department of Agriculture are using this technique to train native bees to pollinate crops that honey bees tend to dominate, as a means to protect our crops against pollinator decline. Blue orchard bees and mason bees are being taught to associate new floral odors with food rewards.

Crafting

Honey bees are extremely versatile, capable of producing honey, wax, royal jelly, propolis, and venom. These products collectively support the life of each individual and contribute to the overall well-being of the colony, as well as benefiting us humans in many ways.

HONEY

A honey bee sucks the nectar out of a flower using her proboscis and stores it in a special stomach-like organ called the honey crop. When a foraging bee returns to the colony, another worker takes the nectar from her honey crop and spreads it over the wax honey comb to facilitate water evaporation. This second bee also adds an important enzyme called invertase to help break down the sugar molecules. Once the nectar has become thick and syrupy honey, it is sealed in a

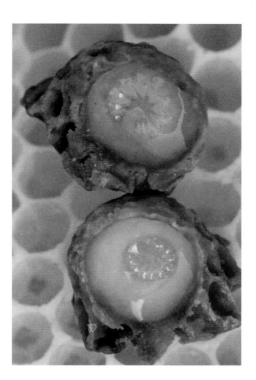

Right *In specially constructed queen cells, two queen larvae are surrounded by royal jelly.*

ROYAL JELLY

Royal jelly is a protein-rich substance made by honey bees. It is secreted by the hypopharyngeal gland, or brood food gland, located on the top of a young worker bee's head. These "nurse" bees feed royal jelly to larvae. More is given to the queen larva, causing her to grow larger than other colony members, to attain sexual maturity, and to live for longer. Royal jelly is made from digested pollen and honey, and it contains sugars, fats, amino acids, vitamins, minerals, and proteins—including royalactin, thought to be a "fountain of youth" protein.

PROPOLIS

Propolis, or bee glue, is a mixture of beeswax and resins collected from leaf buds, twigs, and tree bark. This mixture is used to line nest cavities and brood combs, repair brood combs, seal cracks, and reduce the size of the hive entrance. Propolis also has antibacterial and antifungal properties, and it contains an antioxidant called pinocembrin.

VENOM

Female worker bees of many species produce an apitoxin, or venom, that consists of a complex mixture of proteins. These bees can inject apitoxin into attackers via their sting, a modified ovipositor (egg-laying organ). Venom is stored in a sac beside the sting. Recent research suggests that bee venom may have benefits in human health care. For example, one constituent of venom (mellitin) has been found to destroy HIV-infected cells without killing non-HIV cells.

cell using a wax cap. Once it is sealed, the honey contains so little water that no microbes are able to grow in it.

WAX

Honey bees (and bumble bees) make nests of wax. Beeswax is derived from flower nectar and is composed of esters, or modified fatty acids. Colonies craft wax when they huddle together to raise their body temperatures, and then secrete small flakes from wax-producing glands in their abdomens. These glands only develop fully when bees are between twelve and thirty days old, so older house bees are responsible for all wax production. Once the wax flakes are excreted, other workers take them to different areas of the hive to build brood cells and cells for the storage of pollen and honey.

Nesting

The sizes and locations of bees' nests vary among species, and depend to a great extent on the degree of sociality. Some burrow into the ground to build nests, while others prefer cavities. Solitary bees have small, simple nests, while semi-social and eusocial bees build larger, more elaborate nests.

MINING

Bees belonging to the family Andrenidae are typically mining bees, and nest in underground burrows. Mining bees are solitary and make their burrows in areas of exposed soil with good drainage, such as banks or hills. A female mining bee digs her burrow, stocks the cell with a mixed ball of pollen and nectar, and lays a single egg on top of the ball. Some species lay multiple eggs in the cell and seal the burrow until the mature bees emerge the next spring. Once the larvae hatch, they eat the food and spend the winter in the burrow while maturing into an adult bee.

Right *A newly emerged adult female tawny mining bee* (Andrena fulva) *leaves the nest hole during April in Leicestershire, UK.*

CAVITY DWELLING

There are many species of bee that build their nests in crevices such as hollow plant stems, or holes in wood or masonry. *Anthophora plumipes* is a solitary bee that nests in holes in old walls and may be seen in the castles of Europe and the Middle East. Honey bees and stingless bees both build nests in hollow trees or other cavities. Honey bees build vertical wax combs with hexagonal cells for honey storage, while stingless bees build horizontal brood combs surrounded by large, egg-shaped cells, full of supplies of pollen and honey.

The species *Osmia avosetta*, a solitary bee found in the Middle East, is unique in that it builds cocoon-like brood chambers out of flower petals. A female *O. avosetta* spends two days gathering flower petals, then she sticks them together with nectar, lines the inside with mud, and covers the mud with more flower petals. In the end, each chamber is barely half an inch (1 cm) long, yet it is a remarkable work of architecture.

HOW BIG IS A BEE'S NEST?

Over 90 percent of bee species are solitary. Some common solitary species are mason, plasterer, digger, sweat, and carpenter bees (albeit, sometimes these species have been noted to nest communally, or occasionally even cooperatively). Many solitary bee species build their nests in burrows, and in some cases several female bees will share a burrow entrance but have separate chambers inside. In areas with prime soil for burrowing, many solitary mining bees will nest within the same vicinity, forming a commune. The burrows are typically an inch or so (a few dozen millimeters) deep and half a ¼ inch (6 mm) in diameter.

Below Osmia avosetta *brood chamber, constructed of flower petals and lined with mud.*

Circadian Rhythms

Biological patterns occur in individual bees throughout the day (circadian rhythms), as well as in entire bee societies throughout the year (circannual rhythms, life cycles). Many bee species tend to undergo ingrained behavior that peaks in activity during the day and then troughs at night, a period when many bees engage in sleep-like non-activity.

SLEEP PATTERNS

Bee sleep is described as a period of immobility, typically at night. This may or may not be accompanied by twitching antennae. Could they be dreaming, or are they receiving information from the environment about any potential danger, or the need for activity? During sleep, bees may be standing, or hanging from parts of the hive or from their nest mates.

Societal makeup has an influence. Honey bee sleep patterns change depending on the age of the nest mates surrounding them. Younger bees show stronger rhythms when housed with older bees than when on their own (young bees are poor sleepers). Bumble bee queens have poor rhythms when brood are present, but well-defined ones in the absence of brood (queens sleep better in the absence of developing offspring).

Right *A bee in a typical sleep pose on the underside of a leaf. This is a female* Thygater *sp.*

Researchers have disrupted sleeping bees by attaching magnetic devices to them, and turning the magnet on and off throughout the night. The following day, the waggle dances of these bees were more sluggish and less accurate than those of bees in the same hive that had not had the magnet treatment.

NOCTURNAL FORAGING

Most bees are diurnal, following the typical pattern of flowers—which are open and produce nectar and pollen during the day—but some bees forage during twilight or after sunset. For example, some species of the genus *Perdita*, inhabitants of the desert regions of North America, collect nectar from evening primroses (*Oenothera*). Other bees specialize in collecting food before dawn—as seen in the genus *Xenoglossa*, which forage on *Curcubita* plants in the early morning.

Bees that forage when light levels are low tend to have at least one morphological adaptation aiding this behavior—large eyes (ocelli) that allow more light to enter.

Thermoregulation

Thermoregulation throughout the extremes of winter and summer temperatures is necessary for the survival of bee colonies. For example, thermoregulation is imperative for the year-round survival of honey bee brood, as brood temperatures must remain between 93 and 96°F (34–36°C). Bees have special methods for staying warm through the winter and cooling down in the summer.

HEATING

For many bees, a hairy coat helps to insulate the body, but a large number of solitary species rely on warmth from the sun, since they lack much hair.

Honey bee colonies maintain heat in the nest during colder months, to keep the brood at 95°F (35°C). In the absence of brood, the temperature may drop much lower, but it still remains well above freezing. Bees shiver to keep warm by rapidly contracting their wing muscles. The bees on the outside of the cluster will ensure that the colony stays at a minimum of 41°F (5°C), the minimum temperature at which shivering is still possible. For every 11°F (6°C) drop in temperature they will have to work twice as hard to keep warm and function.

COOLING

Bees have several techniques for protecting their nests against summer heat. Although many species are covered in hair, the ventral side typically remains bare. This allows heat to escape efficiently during hot weather.

Bees that live in colonies often cool the nest by fanning their wings to create airflow. They face away from the hive opening, and push hot air out with their wings. Sometimes, fanning is used in conjunction with evaporative cooling. Evaporative cooling is accomplished by bringing water droplets into the hive and

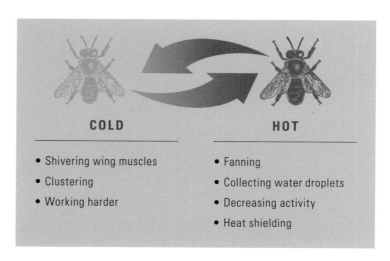

COLD
- Shivering wing muscles
- Clustering
- Working harder

HOT
- Fanning
- Collecting water droplets
- Decreasing activity
- Heat shielding

collectively fanning to create air currents that dissipate the water. Bees also utilize evaporative cooling by flying with their tongues out, allowing water to evaporate and cool the air, not dissimilar to how we humans sweat to cool ourselves down. Western honey bees have been observed to hold droplets of nectar between the mandibles to maintain lower temperatures in the head and thorax.

Heat shielding is another cooling technique, in which worker bees stand on the hottest parts of the nest and absorb the heat in order to shield the temperature-sensitive brood. These individuals trade off jobs when one bee gets too hot, so a cool bee can take over her shift. When a nest overheats to 100°F (38°C) or more, bees can be seen "bearding," or piling up outside the nest to cool off. Once nest temperatures return to optimal levels around nightfall, the colony returns to work as usual inside the nest.

BEHAVIORAL FEVER

The biological definition of a fever is a raising of the temperature above a set-point to clear an infection, and then lowering the temperature back to the set-point once the infection is cleared. Behavioral fever, or social fever, is a variation on this whereby some cold-blooded organisms raise their body temperatures as a group to fight off an infection. US researcher Philip Starks recorded the change in temperature in hives when sugar water with fungal spores was fed to bees and the infection subsequently being cleared. He concluded that honey bees engage in behavioral fever in response to the fungal infection chalk brood disease. The bees vibrate their flight muscles rapidly to generate heat when a fungal infection is present. Interestingly, it has been observed that colonies initiate this behavioral fever even before the infection has killed any larvae.

Changing Behavior Over Time: Temporal Polyethism 🐝

The behavior of a honey bee worker is determined by her age. On emergence, she immediately performs in-hive tasks such as cleaning and helping to feed the developing brood, but as she ages her role shifts through a number of age-related stages. This shift in behavior with time is known as temporal polyethism.

Below *Young worker honey bees (*Apis mellifera*) act as nurse bees, tending the brood.*

YOUNG ADULTS (NURSES)

Newly emerged adult honey bees are often called nurse bees. In their first week of life their focus is on tending the developing brood, and capping off older larvae as they enter pupation. As they enter their second week of life they shift their duties toward attending to the queen's needs and patrolling the hive for any abnormalities, such as genetically abnormal eggs or damaged or dead brood. Much of the nurse bee's activity contributes to the defense of the hive against disease, in the form of aseptic and hygienic behavior.

			Capping brood		Early cell cleaning			Tending brood		Queen tending	Patrolling
Day 1	*Day 2*	*Day 3*	*Day 4*	*Day 5*	*Day 6*	*Day 7*	*Day 8*	*Day 9*	*Day 10*	*Day 11*	*Day 12*

Right *Middle-aged worker bees may cluster at the entrance of the hive in hot weather, a behavior known as "bearding" (see page 77).*

Far right *An older worker specializes in foraging for nectar, pollen, and water.*

MIDDLE-AGED ADULTS (FOOD PROCESSORS & GUARDS)

When they reach two to three weeks of age, worker bees engage in a range of activities that increase their exposure to the outside world, but they remain at the hive. This makes sense in light of their developing immune system. These middle-aged bees receive and process food, taking water out of nectar to make it into honey, and adding enzymes to pollen, enabling beneficial microorganisms to aid in storing it as bee bread. At this age, the bees also act as guards, protecting the entrance of the hive from intruders, and they are responsible for environmental regulation, specifically with regard to temperature.

OLD ADULTS (FORAGERS)

Only the oldest worker honey bees leave the hive to collect nectar, pollen, and water to supply the colony with its daily needs. In most cases, this means that the first foraging trip does not occur until after twenty-three days of life as an adult. The average worker lives for little more than one month during the active summer season when flowers are abundant and there is work to be done, but they may live three or even five times longer in the winter when the hive is dormant. Recent data also suggests that older workers have robust immune function compared to other life stages of bees, presumably to equip them with defense against the pathogens that they are likely to encounter while foraging.

Below The changing role of a worker honey bee. The exact timings are dependent on ecological cues such as a hive's needs.

Receiving nectar Cleaning debris from hive	Resting/ Heat-shielding Late cell cleaning First orientation flight	Comb building	Handling pollen Ventilating		Guard duty					First foraging trip
Day 13	*Day 14*	*Day 15*	*Day 16*	*Day 17*	*Day 18*	*Day 19*	*Day 20*	*Day 21*	*Day 22*	*Day 23*

Defense & Aggression

A bee must defend itself and its nest against a variety of threats, such as robbery, parasitism, and predation. Pathogens can be carried into nests on the bodies of bees or invaders. Enemies come in a wide variety of forms: viruses, bacteria, fungi, arthropods including ants and other bees, birds, amphibians, and mammals, including humans. Various avoidance mechanisms have evolved in response, some in the form of defense, and others of offense.

Below *The European bee-eater (*Merops apiaster*) is a specialist predator on bees and wasps.*

SOCIAL VS. SOLITARY

Social bees, such as stingless bees (e.g., *Trigona*) and honey bees (*Apis*) engage in active group defense of their nests. These bees guard the nest opening and only let related bees in. *Trigona* bees are known to let a robber bee (*Lestrimellita limao*) enter their tubular nest opening, only to be ambushed by the guards, which chew the intruder's wings until it is forced to crawl away. Coordinated group behavior can also warn off predators by looking larger than any individual bee could, as noted in the wave formations of the giant Asian honey bee (*Apis dorsata*). Honey bees regulate defensive behavior by using alarm pheromones.

Solitary bees are limited to passive defenses such as hiding in convoluted or deep nests. Another form of passive defense, again more typical of solitary bees, is shifting activity to nighttime, as seen in the nocturnal-foraging panurgine bees. This avoidance behavior could

Above *The honey badger (*Mellivora capensis*) is a widespread African and Asian species. Its scientific name means "honey-eater."*

have evolved to avoid diurnal predators, or it might be a response to the harsh conditions found in the mainly desert habitats in which they are most common.

BY-PRODUCTS & ADAPTATIONS

Insectivorous birds such as kingbirds and bee-eaters catch bees in flight, carry them to a perch, and then squeeze out the juices of worker bees before finally dropping their bodies to the ground. Perhaps this was a strong selection factor favoring the extreme beats per minute noted in bee flight—an adaptation for adept flight as a defense against aerial predators.

Mammals such as skunks, mice, bears, and honey badgers attack hives and eat bees, their brood, and/or their food stores. The honey badger, in particular, is capable of devastating attacks on bee colonies, and bees and beekeepers can suffer major losses as a result. Beekeepers in Africa often hang their hives from trees in order to avoid predators. The aggressive behavior shown by African honey bees is likely an adaptation for defense against voracious predators, taking the form of bees that are attracted to stings (and sting those marked enemies even more) and also give chase for up to 1,300 feet (400 m).

A HOT, DEFENSIVE BEE BALL

Western honey bees (*Apis mellifera*) are not native to Japan, but are used by beekeepers there as they are better honey producers than the native eastern honey bee (*A. cerana*). Japanese giant hornets (*Vespa mandarinia japonica*) are ruthless predators on the bees' stores of honey and brood. The razor-sharp mandibles and poisonous venom enable one wasp to kill a thousand bees or more per attack. Remarkably, Japanese honey bees (*A. cerana japonica*) are able to fight back by utilizing their group defenses, which have likely evolved through natural selection in response to Japanese hornets obliterating hives with weak defenses. The key strategy is heat. A group of hundreds of bees surrounds a hornet, and by contracting their wing muscles rapidly they generate enough heat to kill the hornet at 117°F (47°C), a point beyond which bees can survive but the predators cannot.

Parenting ✦

Bee species that are eusocial, communal, or solitary differ in their parenting techniques. In eusocial species, the queen's only task is to lay her eggs, and the workers care for them. Most solitary species lay their eggs, leave a food provision, and never return.

NO PARENTING

In most eusocial bees there is no brood care performed by the egg-laying individual. Eusocial bees, such as honey bees, live in complex groups with overlapping generations, reproductive division of labor, and cooperative rearing of brood (larvae and pupae). These eusocial bee species usually consist of one egg-laying queen with a team of worker bees to care for the brood. The queen only lays eggs, and does nothing more for them directly. It is the worker bees, and not the parent, that are responsible for feeding the larvae, keeping them clean, sealing larval cells for maturation, and even maintaining the temperature of the larvae as they develop.

  Cuckoo bees, in the subfamily Nomadinae and in the family Apidae, avoid parenting duties by laying their eggs in egg cells within the nests of other bees. These cells are full of food left by the

Below *Bumble bee nest.*

Left *An adult female cuckoo bee (*Stelis phaeoptera*) breaks into the nest of a mason bee (*Osmia leaiana*).*

Right *Bee larvae (Ceratina sp.) feed on a pollen mass that has been placed in the nest by the female prior to egg laying.*

host bee for her own young, so the cuckoo bee does not even need to forage food for her offspring. This behavior comes at a cost, however, because the host bee often discovers and eats the cuckoo larvae.

MINIMAL PARENTING

With solitary bees, parenting involves provisioning the nest with a ball of pollen, laying the eggs, and leaving. Mining bee species, for example, practice extensive parenting of this type, with the female providing only a small amount of food. Similarly, a carpenter bee digs a tunnel into a piece of wood, constructs a ball of bee bread consisting of nectar and pollen, and lays her eggs on top. She then seals the tunnel with sawdust chips and leaves only a small chamber where her eggs are. These eggs are left with no more than the bee bread provision, and the bees are on their own to dig their way out of the hive once they emerge from their cells.

INTENSIVE PARENTING

In bumble bees species, the queen initially directly cares for her eggs after laying them. Although bumble bees are communal, a new queen is responsible for starting her colony each year. She lays her first batch of eggs and forages for food, providing for and incubating her young developing brood. A bumble bee queen tends to lay her eggs in batches of five to ten on balls of pollen that she has created. She then seals her brood cells with wax and incubates them. When the first females hatch, they become the worker bees and take over the foraging and provisioning duties while the queen lays and incubates her next batch of eggs.

Courtship 🐝

In most solitary bee species, a female will only mate once. This means that there is intense competition among males to mate with a female. The queens of social species often take just a single mating flight in their lives—but they will mate with several males. While courtship behavior still remains for the most part unknown, there are some consistent patterns that have been observed, distinguishing solitary from social bee species.

Below *A mating ball of digger bees (*Diadasia australis*) Pawnee National Grassland, Colorado.*

Right *A male orchid bee* (Euglossa ignita).

SOLITARY BEES

Solitary bee species utilize several different strategies in mating behavior to find mates. Male bees are usually born first each season, and then wait for the females to emerge. In some species, such as the sweat bee (*Nomia triangulifera*), the males will fly near the females' primary food source and utilize muscles in their forelegs, midlegs, metasoma (abdomen), antennae, and indirect flight muscles, to produce loud buzzing sounds that will attract the females.

In some species, males search for females alone, but others search together in swarms. Competition is fierce. As a result, males of some species will set up a territory by marking flowers with pheromones. These territories can range in size from 5 to 65 square feet (0.5–6 m²). Other species will deposit pheromones on females during copulation to deter other males from mating.

The solitary orchid bee males gather fragrances from orchid flowers and other plants to store in pockets on their legs. These fragrances are released at courtship territories to attract females. Since many males are present at courtship territories, the female may be able to measure the fitness of each of the males from these fragrances, and use them to choose her single mate.

SOCIAL BEES

In social species of bee, the queen has a special pheromone—the queen mandibular pheromone (QMP)—that serves several functions, including the attraction of drones during their mating flights. Female bees of social species such as honey bees mate with many males in the span of one mating flight. Queens tend to fly to areas above landmarks, such as boulders or church steeples, to find areas where drones congregate.

A queen honey bee may make as many as three mating flights, mating with an average of twelve drones in all, but genetic data have revealed evidence for up to twenty-nine genetically distinct males mating with one female. Drones usually gather at the nearest drone congregation area and wait for the queen to arrive. Male honey bees die during copulation, and so the competition with other males over mating with high-quality queens is likely fierce, albeit not yet well understood.

Sexual Reproduction

Sexual reproduction is the process of combining genetic material (DNA) from two individuals to create new organisms. Charles Darwin marveled at the evolutionary forces that produced sex in his 1871 book, *The Descent of Man, and Selection in Relation to Sex.* Darwin pondered the paradox of traits that should not be favored by natural selection, but that existed nonetheless, such as the peacock's tail, the deer's antlers, the lion's mane, and indeed the social bee's sterile workers.

INTER-SEXUAL COMPETITION (MALE VS. FEMALE)

Inter-sexual competition is a battle between the sexes. For a male, sperm are cheap, and he passes millions to the female in one sexual exchange. Eggs are more costly, requiring more nutritional, energetic, and temporal resources to produce. In bees as in other animals, therefore, females are typically more selective than males when choosing a mate—and the males have to work hard to woo a female. Some bee species show elaborate courtship displays by drones, as noted in red mason bees (*Osmia bicornis*), carpenter bees (*Xylocopa* spp.), a solitary halictid bee (*Nomia triangulifera*), and others.

Left *The young queen honey bee (*Apis mellifera*), hatched five or six days earlier, typically mates with several males during her mating flight.*

INTRA-SEXUAL COMPETITION

Intra-sexual competition arises when individuals of the same sex compete with each other for access to mates. Within the bee world, conflict between males occurs when drones compete for access to queens for mating, and at certain times in the life cycle of the colony there is also conflict between females that takes the form of fierce competition when more than one queen has been reared inside the nest.

Most species of bumble bee are monandrous, meaning that mating takes place between a single female and a single male, with no multiple mating. Because of the competition that exists among male bumble bees to mate with the female, the male has a special "mating plug" that is left inside the female after mating. This plug seals off the reproductive tract and prevents the queen from mating with other males, thereby ensuring that the one male who did mate with her successfully passes his genes on to the next generation. This evolutionary adaptation is especially important for the male, because he dies immediately after copulation as the endophallus is ripped out and left in the female's body.

RESOURCE-DEFENSE POLYGYNY

Unlike most male bees, the male wool carder bee (*Anthidium manicatum*) dominates the females. These males defend territory, typically a bush or flower patch—any habitat with nesting or foraging value—and guard it ferociously. They take swirling patrol flights, and nothing is allowed access except for female wool carder bees. While females forage for nectar and pollen within his territory, the male gains access to his harem. Should any other male enter his territory, he dive-bombs the intruder, clipping the invader's wings with sharp barbs on the front of his forelegs. The shredded wings prevent the victim from flying, and he falls to the ground, left to see his own way out of the patch on foot. Male wool carder bees are even known to dive-bomb other insects.

Other bees also engage in territory defense. For example, the hairy-footed flower bee, otherwise known as the plume-legged bee (*Anthophora plumipes*), guards his territory by patrolling his nesting site and foraging resources, and using his bulky head to ram intruders in flight. But no bee is as determined as the wool carder bee, which displays the ultimate in resource-defense polygyny.

Mating Systems ❧

Unlike most of the natural world, bee societies are female-driven, with limited exceptions. Their mating systems reflect this, with very few apparent examples of polygamy (one male, multiple females). Bee mating systems can generally be characterized as monandrous or polyandrous. Monandrous means that a female only mates with one male in her lifetime. Polyandrous means that a female mates with multiple males throughout her life. Furthermore, hyperpolyandry is when one female mates with many males, typically over a dozen or so.

These different mating systems provide different genetic advantages to the species characterized by them. Polyandrous mating assists in the continuity of the colony for some species, while monandrous mating can ensure that the strongest male genes are passed on to support colony cooperation.

MONANDRY

When a bee species is monandrous, all the eggs that are destined to develop into workers are fertilized with the sperm of a single male. This high level of relatedness may lead to high levels of cooperation in the colony, and thus improve productivity and reproductive fitness. One might therefore predict that social bees would favor a monandrous mating system, so as to increase cohesion among the workers. There are indeed several species of eusocial bees that are characterized by monandrous mating habits. For example, many stingless bees are eusocial, and a queen will only choose one male to mate with, as predicted. Monandry is also typical of many species of bumble bee.

POLYANDRY

Polyandrous bees have mating systems that involve multiply mated queens. These queens often store the sperm from all of these males for their lifetime, within the highly convoluted spermatheca. In polyandrous species, the queen leaves the hive on a mating flight, typically within one week of emerging as an adult. Multiple mating allows the queen to lay eggs that are genetically diverse, which may help to prevent inbreeding, and thus the production of genetic misfits (see pages 34–35).

HYPERPOLYANDRY

The mating system of western honey bee queens is near the upper limit of

Above *Mating carpenter bees (*Xylocopa *sp.).*

hyperpolyandry noted in the natural world. The success of *Apis mellifera* suggests that the benefits of hyperpolyandry outweigh its costs. On the one hand, potential benefits include improved disease resistance, defense against enemies, diversity of beneficial bacteria, and efficiency in hygienic behavior. On the other hand, potential costs include sexually transmitted diseases, decreased genetic relatedness among workers, greater wear and tear on anatomy, and intra-sexual competition. In one hive in Massachusetts, USA, it was shown that a single female laid eggs fertilized by as many as twenty-nine different fathers.

Left *Hyperpolyandry: one female, many males.*

CHAPTER 4

Bees & Humans

Ancient & Modern Knowledge

Whether we are considering the wisdom of the ancients or the work of twenty-first-century scientists, the enquiring human mind has two primary foci: basic and applied research. Basic research is interested in advancing human understanding of how the natural world works, for its own sake. Applied studies look at ways we can advance the human experience. Bees were, and still are, an integral resource in both types of study.

BEE SCIENCE IN THE ANCIENT WORLD

The bee was first studied by the ancient Egyptians—as shown by illustrations of bees that can be seen in wall paintings from the time. Honey and wax were used throughout the eastern Mediterranean and beyond as a means of preservation. The Syrians, Babylonians, and Assyrians shared this interest.

Honey and wax preserve food because of their hygroscopic (drying) properties, their anaerobic nature, and their high sugar content. Ancient societies likely used bee products as a means to lengthen the shelf life of food. Additionally, human mummification involved coating a corpse with a mixture of honey, wax, and propolis from beehives.

In the fourth century BCE, Hippocrates and Aristotle both commented on the benefits of bee science to humanity, as did the Roman scientist Pliny the Elder some four hundred years later.

Left *Bees on a honey comb. An image from the 1623 edition of Charles Butler's* The Feminine Monarchie: or, the Historie of Bees.

THE RENAISSANCE & AFTER

Ever since the Renaissance, natural historians have been fascinated by bees, and there are several early printed books about them. One of the best known was written by the Reverend Dr. Charles Butler (1559–1647), whose *The Feminine Monarchie* was published in 1609. Although not the first to realize this, Butler was the first to record in print that the large bee in a colony of honey bees is female, and therefore a queen rather than a king—as was often thought at the time. Butler also described scout bees or "spies" in swarms, although their role was not fully understood until recently.

Another clergyman, the Reverend John Thorley (1671–1759) published his *Melisselogia, or the Female Monarchy* in 1744. He realized that beeswax is secreted, not collected from plants as had been thought, noted that queens lay eggs and described their "piping." He also realized that bees can recognize colors.

In the nineteenth century, Charles Darwin (1809–1882) made many observations about bees and their interactions with plants. For example, he made a study of different orchid species and the ways that they are adapted to different kinds of insect pollination.

In later years, bee research was no longer just the preserve of wealthy individuals, as state funding became available. In the mid-twentieth century, a good example is the work of Dr. Colin Butler at Rothamsted Experimental Station in the UK, who first realized that queen honey bees produce a substance, now known as queen mandibular pheromone (QMP), that controls the organization of a colony.

Scientific Research Today 🐝

The biologist Karl von Frisch wrote that "The bee's life is like a magic well: the more you draw from it, the more it fills with water." Bees are thus a valuable resource for scientific research, and they have contributions to make in both basic and applied studies.

The value of bees as pollinators has long been recognized, but research has often focused on honey production. In the USA, the National Honey Board (NHB) was formed after the enactment of the Honey Research, Promotion and Consumer Information Act in 1984. In recent years, losses of honey bees and the phenomenon of colony collapse disorder have led the NHB to broaden its research grant program beyond a focus on honey to include wider aspects of honey bee research. Scientific research continues to expand today, with an ever wider range of bee species coming under the microscope.

BASIC RESEARCH

Bees continue to advance our understanding of seemingly unrelated topics, including aging, learning, memory, epidemiology, sociology, and more. Charles Michener provided tremendous

advances to bee classification, beginning this work in the 1950s and refining the system in the 1990s. In 2000, Michener's book *The Bees of the World* became the leading authority on bee evolution, systematics, and identification.

In the twenty-first century, researchers have focused further on bee biology (examining topics such as hygienic and aseptic behavior, Africanized bee behavior, and queen mating), and they have also turned their attention to ways that bees can be used to test general hypotheses related to other aspects of biology. Bees have made significant contributions to studies of social behavior, aging, thermoregulation, immunology, epidemiology, and even economics.

Above *The future of pollination? Robobee, shown here alongside a ⅜-inch (1-cm) bolt, is being developed at the Harvard School of Engineering and Applied Sciences.*

APPLIED RESEARCH

How can bees advance the human condition? Past research has developed a basic understanding of their pollination processes, their various products, and their economic contribution, and the frontiers of knowledge continue to be pushed forward.

For instance, the US Defense Advanced Research Projects Agency (DARPA) recently funded research projects to investigate the use of bees for tracking land mines.

DARPA has also funded a project at the Harvard School of Engineering and Applied Sciences to develop a robotic bee. This bee senses cues from the environment, with the aim of helping us to understand complex biological systems, but the researchers speculate that their RoboBee could even have a practical application as a pollinator of crops, should bee populations not recover from their current challenges.

Below *In an experiment in New Mexico, the ability of honey bees to detect land mines is tested. The anti-tank mines shown here are unfused, and therefore safe for both bees and humans.*

An Economic Force

The economic contribution of bees to humanity is huge, yet difficult to quantify. In the year 2000, honey bees alone were estimated to contribute $14.6 billion to the US economy, and the worldwide figure is something like €153 billion ($207 billion).

At least 130 fruit and vegetable crops depend on insects for pollination. However, increasing demand for honey bee pollination services is outpacing the availability of bee colonies. The number of honey bee colonies in the USA has decreased dramatically from 4.5 million in the 1940s to 2.5 million today. The rate of decline varies across the globe, but the greatest declines in honey bee populations have been seen in the northern hemisphere.

BEES AS POLLINATORS

Bees are major players on the global economic stage, mainly because of their role as pollinators. An astonishing range of fruit and vegetable crops are 90 percent or more reliant on insects for pollination. The yield of these crops would decline to less than 10 percent of its current level, if bees disappeared.

100 PERCENT DEPENDENT UPON INSECT POLLINATION		90 PERCENT DEPENDENT UPON INSECT POLLINATION
• Alfalfa	• Carrot	• Cherry
• Almond	• Cauliflower	• Cucumber
• Apple	• Celery	• Kiwifruit
• Asparagus	• Cranberry	• Macadamia
• Avocado	• Legume seed	• Pumpkin
• Blueberry	• Onion	• Squash
• Broccoli		

Many crops are entirely dependent on bees and other insects for pollination.

Far left *Pumpkin*
(Cucurbita *sp.*)
Left *Almond*
(Prunus amygdalus)

Far left *Apple*
(Malus domestica)
Left *Cucumber*
(Cucumis sativus)

Far left *Onion*
(Allium cepa)
Left *Blueberry*
(Vaccinium *sp.*)

BEES AS PRODUCERS

Honey is a valuable food resource throughout the world, with a relatively stable average retail price hovering around $5 per pound ($11 per kg) in the USA between 2006 and 2013, according to the National Honey Board. Bulk honey imports into the USA have, however, increased sharply. Bulk prices were around $1.50 per pound across 2011–2012, but the first five months of 2013 were notably more expensive for bulk honey buyers, averaging $1.70 per pound in the first five months—a 13 percent increase.

Wax is also a highly valuable commodity on the global market. It can even be more valuable than honey, because it can be used for a wide variety of different purposes. Beeswax acts as a preservative for fresh foods, such as apples and pears. The cosmetics industry accounts for an estimated 40 percent of its use, and other high-volume uses include within the pharmaceutical industry (30 percent of total use) and for candle making (20 percent). Wax is also used in the manufacture of lubricants and polishes, in crayons and encaustic art, for strengthening threads, in electronics, and in a host of other applications.

Spirituality

Bees and their products have a long association with human religions and rituals. Honey has been used throughout human history to convey blessings, as well as being used as an offering to the gods and spirits, and to provide protection in this life and the next.

BIRTH, MARRIAGE & DEATH

Many cultures have traditions built upon products obtained from bees, especially honey. In Slovakia and elsewhere in eastern Europe, for example, families eat a thin wafer on Christmas Eve. Called oplatky, it is imprinted with a religious scene. On this is spread honey with slivers of sliced garlic. The garlic is for boosting the immune system while the honey sweetens the bitterness and is a symbol for a sweet year.

A child, on the first day of learning the Torah, licks honey off a Hebrew letter so that all future learning will be associated with sweetness.

Many cultures use wax and honey in both marriage and death rituals. In the Middle East, a bride and groom feed one another honey and lemon from ornate vessels to signify accepting the sour with the sweet.

Egyptians, Babylonians, Persians, Assyrians, and Arabs all used wax and honey to embalm their dead.

In joy and sadness, for thanks and for requests for help, the bee has long been an integral part of human life.

SHAMANISM

Bees played a role in goddess worship and shamanism as well. Artemis (also known as Diana), the many-breasted goddess of fertility, was often depicted with beehives. The priests and priestesses in her temple were called bees, and the entire community, the hive. Melissae, relating to the honey bee (*Apis mellifera*), was also the name of Aphrodite's priestesses.

Above *Seventh-century BCE gold plaque depicting a bee woman or bee goddess, perhaps associated with the Greek goddess Artemis.*

The genus *Melipona*, a stingless honeybee revered by Mayan shamans, was celebrated in a ritual ceremony twice a year. And if a bee were accidentally killed during honey collection, the shaman would bury the body of the bee in the ground after wrapping it in a leaf.

SPIRITUALITY

Even without formal ritual, beekeeping can be a spiritual experience. Watching the inner world of bees, especially in an observation hive, can be relaxing and calming. Entering another world, forgetting momentarily the troubles of one's everyday life, can bring one into a different state of awareness.

When a beekeeper dies there is a tradition, at least in the USA, that the hive is covered with a black cloth. For one day the bees remain in the hive and "in mourning." The lives of the beekeeper and the bees are intimately intertwined, and this ritual marks the loss of that connection between human and bee.

BULGARIAN BEEKEEPING HOLIDAY: SAINT PROCOPIUS BEEKEEPERS' DAY

Cultures that recognize and respect bees have created rituals to honor them. In Bulgaria, July 8th is designated as the beekeepers' holiday in honor of Saint Procopius (or Prokopi). One Bulgarian bee ceremony is performed in the open fields beside the beehives. Six women (Babis), representing the hexagonal shape of a beehive cell, stand around a woman who represents the Mother Queen. A song to the bees is sung called "Pchelice Medna Oliadina," in praise of the sweet, small honey bee.

Islam, Judaism & Christianity

Bees and beekeeping make numerous appearances in the sacred texts of many of the world's religions.

ISLAM

In the Muslim faith, basic tenets of hard work, loyalty, and devotion are written metaphorically using imagery of the bee. Merchants selling honeys from far and wide are a relatively unusual part of Islamic culture. The value of honey as a healing agent to the body may be linked to the power of the Qur'an as a healing agent to the mind:

And your Lord inspired to the Bee, saying:
"Take your habitations among the mountains, houses, and trees
Then eat of all the fruits, and follow the ways of your Lord, laid down (for you)."
There emerges from their bellies a drink of varying color
In which there is healing for people. Verily, in this is indeed a sign for people who think.

QUR'AN, SURAT AN-NAHL (THE BEE) 16:68–69

Left Samson Offering His Parents a Honeycomb *(oil on canvas) by Guercino (Giovanni Francesco Barbieri) (1591–1666).*

CHRISTIANITY

In the Christian tradition, many monks were beekeepers—including Gregor Mendel, the father of genetics—and bees became the symbol for the most excellent type of servant, the kind that humans should emulate, being unified in their work for the good of all without complaint or concern for hierarchical position. Saint Ambrose saw the hive as a representation of the monastic life. Saint Francis de Sales used the bees as an example of how to conduct the human spiritual life:

> Watch a bee hovering over the mountain thyme;—the juices it gathers are bitter, but the bee turns them all to honey,—and so tells the worldling, that though the devout soul finds bitter herbs along its path of devotion, they are all turned to sweetness and pleasantness as it treads.

INTRODUCTION TO THE DEVOUT LIFE,
SAINT FRANCIS DE SALES (1609)

Above *The prophetess Devorah singing a victory hymn in triumph over the Canaanites.*

JUDAISM

The sacred texts of the Torah, Midrash, and Talmud all use bees as metaphors for looking to a leader for guidance, obedience to teachings, and accumulation of good things for selfless purposes. Israel, referred to as the land flowing with milk and honey in the Bible, is a prosperous home. The name of the prophetess Devorah (Deborah), who led the Jewish people from 2654 to 2694 in the Jewish calendar (around the twelfth to eleventh centuries BCE), means bee in Hebrew. Today, honey is incorporated into the Rosh Hashanah celebration of the New Year.

> I shall rescue them from the hand of Egypt and bring them up to … a land flowing with milk and honey.

THE BIBLE, EXODUS 3:8

Right *Gregor Mendel (1822–84)—famous for his work on the genetics of peas, and as a beekeeper.*

GREGOR MENDEL
Abbot of Brünn
Born 1822. Died 1884.
From a photograph kindly supplied by the Very Rev. Dr Janeischek, the present Abbot.

Patron Saints of Beekeeping

*May thy holy blessing descend upon these bees
and these hives,
so that they may multiply, be fruitful and be
preserved from all ills
and that the fruits coming forth from them may
be distributed for thy praise
and that of thy Son and the holy Spirit and of
the most blessed Virgin Mary.*

"PRAYER FOR ST. BENEDICT'S DAY, MARCH 21ST,"
FROM *A CANDLE IS LIGHTED*, BY P. STEWART CRAIG,
EASTCOTE, MIDDLESEX (1945)

In Roman Catholicism there are many saints who are associated with or symbolized by bees or beekeeping. Saints are revered on their respective feast days each year.

St. Benedict's prayer (above) addresses the patron saint of beekeeping. Some devout beekeepers in Europe, especially in France, hang his image from their hives, impressed on small metal pendants.

St. Ambrose was found as a boy by his father with his face covered in bees. Throughout his life, many described his preaching to be as sweet as honey. He is a patron saint of beekeepers, bees, and candle makers. St. Ambrose is often symbolically depicted with a beehive or bees, which also represent wisdom.

Below *St. Valentine, the patron saint of beekeepers as well as of love, is shown in a hand-colored fifteenth-century German woodcut.*

Right *St. Benedict, whose image is often still seen on beehives to this day.*

Far right *St. Bernard of Clairvaux, depicted here in the white habit of the Cistercian Order, which he joined in the early twelfth century.*

St. **Valentine** is another patron saint of beekeepers as well as the patron of love, young people, happy marriages, and against epilepsy. Although his connection to bees is not understood, it may be because the sweetness of honey is metaphorically related to the sweetness of love.

St. **Gregory's** book of animals mentions the bee as a symbol of sweetness, to represent a community working and thriving in harmony. St. Gregory is associated with flowers opening for the bees in the springtime.

St. **Bartholomew** is primarily noted as the patron saint of tanning. His annual celebration feast occurs in late August, coinciding with the honey harvest—hence his association with beekeeping.

St. **Kharlamii** emphasized natural methods of healing the body's ailments, notably through the use of honey and beeswax. His teaching spread throughout the region south of the Black Sea, spanning Asia and Europe, leaving a legacy of keeping bee products at the altar.

St. **Gobnait** is a patron saint of bees and beekeepers in Ireland. Legend says that she protected her parish from the theft of their cattle by unleashing a swarm of bees to attack the thieves. It is believed that she had a very close relationship with her bees, and she utilized honey in treating illness and wounds.

St. **Modomnoc** (Dominic), another patron saint of bees, may have been the first to introduce honey bees to Ireland after studying bees at a monastery in

Wales in the sixth century. He built a strong relationship with the bees and loved working with them and talking to them. Stories say that when he had finished studies the bees would not allow him to return home to Ireland without them, and they all swarmed to his boat. The bees refused to be left behind, so he brought them home to Ireland with him and continued to be their keeper.

St. **Bernard** is a patron of beekeepers, bees, candle makers, wax melters, and wax refiners. Much like St. Ambrose, St. Bernard is associated with bees because of the sweetness in his rhetoric. St. Bernard is often depicted with a swarm of bees, and is sometimes symbolized by a beehive or bees.

Political Symbolism

Bees, and honey bees in particular, have been valued as a representation of monarchist societies since ancient times, but little did people realize that they are not really monarchic organizations. In truth, the honey bee may better reflect a democratic form of government. Queen honey bees "lead" their colonies via their pheromones, which keep the colony working and suppress egg-laying abilities in worker bees, but it is the workers that hold the decision-making power, especially in terms of choosing foraging and nesting sites.

Above *Pope Urban VIII (1568–1644), here commemorated on a plaque in Florence, Italy, was a member of the Barberini family, whose coat of arms showed three bees.*

MONARCHY

Bees have long been used to symbolize monarchies, especially when that was the prevalent form of government, and at a time when the queen bee was thought to be male.

In early modern history, the beehive was often used in allegory to identify with human society. The queen was praised as a strong leader who commanded her loyal community—who were thought to obediently work for the greater good of the colony. Many rulers incorporated bees into their royal symbols. Napoleon, the self-crowned first emperor of France, represented his dynasty with a bee.

DEMOCRACY

The work of US biologist Tom Seeley has revealed that beehives are not simply a monarchy in which one queen rules over obedient workers. Seeley outlined these ideas in his book *Honeybee Democracy* (2010). When studying swarm behavior, he determined that a honey bee swarm chooses a new place to nest in a democratic fashion. The queen merely accompanies the swarm to the new nesting site, while the rest of the colony makes the decision.

The nesting site is decided upon when the swarm collectively scouts out potential locations, debates over which location is the best, and builds a consensus to make the final decision. This is analogous

Left *The Emperor Napoleon 1 in his coronation robes— adorned with images of bees.*

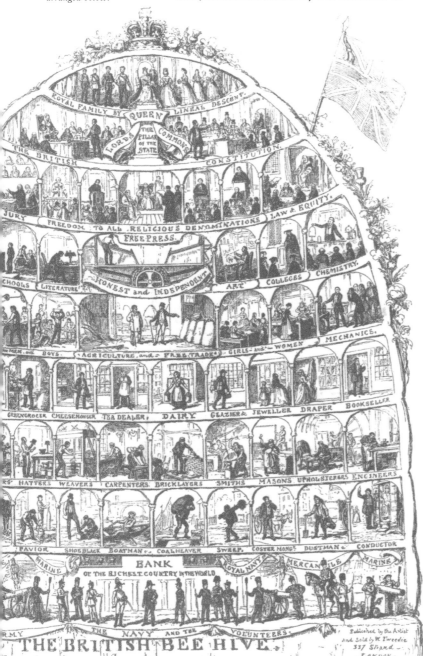

Below *George Cruikshank's* The British Bee Hive, *showed the queen at the top of mid-nineteenth century society and various institutions and trades hierarchically arranged below.*

to a human democracy in which several candidates are chosen, debates occur to allow voters to understand the values of each candidate, and finally a vote takes place to determine which candidate has the most support. In the case of the honey bees, individual scouts fly out in search of a new nesting site, and return to the hive to report back through the medium of dance. Multiple bees dance within a short period of time to convince the nest mates that their site is the best quality. New scout bees are convinced by the vigor of the dances—in effect, a debate on the dance floor—and then fly out to confirm. Eventually, a nest site is decided upon based on the majority of bees dancing vigorously for the winning site.

Overall, Seeley points out, decision-making groups are collectively more intelligent than the smartest individuals within the group. Although the queen is superior, all decision making relies on the colony building a consensus.

ANARCHY

Despite the appearance of organization in the honey bee colony, anarchy is never far away. If the queen is lost by whatever means, workers will try to rear a replacement from young larvae by feeding extra rations of royal jelly. If this fails, in the absence of queen pheromone, the ovaries of workers begin to develop and they may lay eggs. Being unfertilized, these eggs become drones, which only creates more mouths to feed. Drones play little or no part in colony functioning, and so excess drone production is a waste of energy for the colony.

Without a queen to lead them, the entire colony will fail and die. UK biologist Francis Ratnieks has, however, shown that even in a colony with a laying queen, some workers may show ovary development and begin to lay eggs. Other workers "police" the hive and normally eat these worker-laid eggs. If this policing breaks down, anarchy will result.

What if Bees Were to Disappear?

It is now believed that it is unlikely Einstein said that we would all die out within four years if the bee became extinct. But over the generations our diets have shifted from wind-pollinated grain crops to insect-pollinated fruit and vegetable crops, and we have become ever more dependent on bees for nutritious food and a varied diet. This is a problem, at a time when the bees are struggling to keep up with our intensive agricultural practices. And all over the world, native bee species are disappearing from their former ranges, in large part due to habitat loss. What if such losses were to continue? Might we starve? How else might the disappearance of bees affect the lives of humans across the world?

IMPACT ON FOOD PRODUCTION

An estimated 35 percent of global food supplies are either reliant on or are improved by animal pollinators.

Wind-pollinated crops, such as cereals, manage without bees, so we would still have our daily bread. Crops such as oilseed rape do not rely solely upon bees for pollination, but bee pollination significantly improves the quantity and yield, so the loss of bees would also result in less rape oil being available.

If bees disappeared tomorrow, given human adaptability, we'd get used to our bland diet and find other things to eat—eventually. And yet the loss of extra

Below *Agricultural changes often lead to declines in bee numbers, which can lead to the loss of flower-rich meadows.*

Above *Hand pollination of pear trees in China.*

income enjoyed by farmers for producing high-value insect-pollinated crops might mean they could no longer produce the wind-pollinated staples of our diet. Or shortages might mean that food would become very expensive. Either way, this would be bad news for us.

IMPACT ON ANIMAL PRODUCTION

The production of animal fodder would also suffer if bees were lost. Historically, we had clover-rich meadows, pollinated by bees, which were used to raise and fatten sheep and cattle, giving us the expression "living in clover." Alfalfa, related to clover but much taller, remains an important animal fodder worldwide. Cut and baled as hay for horses and cattle, it is also fermented and fed as haylage to dairy cows. In 2009, some 30 million hectares was grown. The USA, as the world's largest alfalfa producer, cultivated 9 million hectares, with the crop pollinated by the non-native alfalfa leafcutter bee (*Megachile rotundata*) rather than the western honey bee. These crops help us to produce more meat and milk. Without bees, we could feed our farm animals with grass, oats, and corn—but their winter feed would be much poorer.

CHAPTER 5

Beekeeping

The Basics 🐝

Beekeeping is an amazing experience. The connection between humans and honey bees is deep-rooted in our history together, spanning tens of thousands of years. Honey bees are very much like people—they go out during the day, come home at night, store food, and interact with family. Watching bees coming and going from the hive can be relaxing and enjoyable in itself, but some people find they want to go a step further, opening up the hive to discover the world within that is revealed only to the beekeeper.

THE GOLDEN RULE

One of the most common sayings in beekeeping goes, "Ask ten beekeepers a question, and receive eleven different answers." Above all else, the beekeeper must always have fun. The Golden Rule of beekeeping is therefore exactly this—if you're not having fun with it, then you're not doing it right. Don't get caught up with the fact that bees sting, and don't let the cloud of doubt and inexperience overwhelm you. Channel those nerves into bravery, take a few deep breaths, and simply observe the bees. Enjoy their behavior, and let your instincts guide you. Take care to observe all you can about typical bee behavior and healthy hive appearance. These observations will form the fundamental core of all your future beekeeping, for anything abnormal will thereafter be easy to spot.

Above A beekeeper has privileged access to the intriguing private life of bees.

EDUCATION

The best way to learn is to connect with other beekeepers. There are countless beekeeping schools and clubs around the world, readily found with a few strokes on the keyboard. Connect with local beekeepers to learn about sources for the best-quality honey bees and equipment. Group learning allows beekeepers the opportunity to learn hands-on, to discuss whether what one person sees is typical or not, and to practice what to do when the abnormal appears. When something unusual is spotted, do some research within your local area. Find local beekeeping texts, ask neighborhood beekeepers, and find government recommendations. Diseases, pests, and environments vary, and so do local beekeeping practices.

Right *A bee sting can be painful, but more serious reactions are relatively rare.*

BEE STINGS & ALLERGIC REACTIONS

Beyond having fun, safety is paramount for the beekeeper. Most honey bees are not aggressive, but beekeepers do get stung from time to time. Everyone's immune system is different, and can also change over time—some reactions diminish to nearly nothing, while others can get worse—and so understanding one's personal reaction to stings is important. Reactions occur either locally or systemically. Local reactions typically involve swelling, redness, and/or itching at the sting site, while systemic reactions involve symptoms beyond the local sting. Severe reactions are rare, but they can involve anaphylaxis (a rapid-onset allergic reaction that often leads to an inability

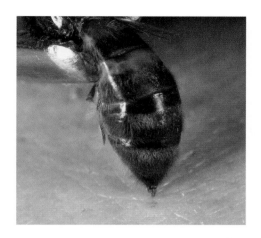

to breathe), and this can be fatal. Those unlucky enough to suffer severe reactions should not keep bees, or should ensure that they have access to epinephrine (adrenaline), which typically requires a doctor's prescription.

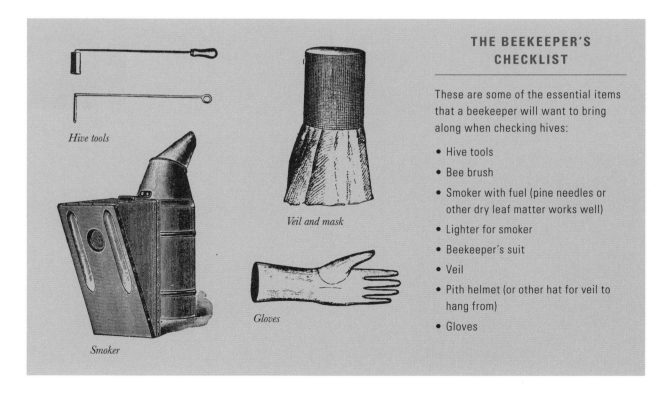

Hive tools

Veil and mask

Gloves

Smoker

THE BEEKEEPER'S CHECKLIST

These are some of the essential items that a beekeeper will want to bring along when checking hives:

- Hive tools
- Bee brush
- Smoker with fuel (pine needles or other dry leaf matter works well)
- Lighter for smoker
- Beekeeper's suit
- Veil
- Pith helmet (or other hat for veil to hang from)
- Gloves

Keeping Other Bees 🐝

Honey bees (of the genus *Apis*) are the most commonly managed bee species, but they are not the only ones. Stingless bees, bumble bees, carpenter bees, wool carder bees, mason bees, and orchard bees are all amenable to management by humans. Knowing about the bees in one's local environment facilitates much of this practice, as the key to managing native bees is to create an enticing nesting space, which the bees will choose as an ideal place for their new home.

COSTS & BENEFITS

The financial cost of keeping native bees is far lower than that of keeping honey bees. Many of their nests can be found in fallen trees or branches, and then either visited where they are or transported to one's property within the log, as is often done with stingless bees in Australia. Existing nests of native bees can be transferred to larger spaces, although this does most often increase the risk of bee mortality. Stingless bees can be kept in box hives similar to Langstroth hives, while bumble bees can be kept in simple shoe boxes.

The benefits of hosting native bees extend well beyond the personal enjoyment to be gained from watching the bees coming and going. Bees other than *Apis* are also vitally important pollinators of local flora, and may be better suited than honey bees to pollinating crop plants. A few species make honey and wax, albeit in smaller quantities than honey bees, but stingless bee honey, for example, commands a premium price.

TRAP NESTS

Most bees can be acquired by setting up trap nests—potential nesting sites that bees will move into and create a nest within. Crevice type and height above the ground are both important

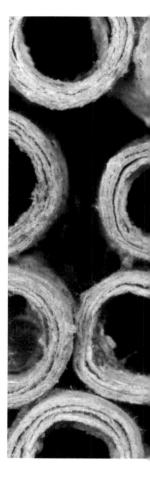

Above *A cardboard tube provides a perfect nesting site for this adult female red mason bee (*Osmia bicornis*).*

Left *A mason bee, (*Osmia cornuta*), makes use of a hole drilled in a log.*

Stingless bees *(Melipona)*
Range: Found in South America, Africa, and other tropical regions.
Method: Keep in the original log, or move to a multi-box hive.

Bumble bees *(Bombus)*
Range: All continents except Australia and Antarctica.
Method: Using gloved hands or a tool such as a shovel, gently transfer an existing nest into a container the size of a shoe box. Bumble bee nests may be kept indoors—for example, for greenhouse pollination. The box should be on or near the ground, and filled with fluffy material such as cotton, with a small hole (about as wide as a thumb) on at least one side for an entrance.

Carpenter bees *(Xylocopa)*
Range: All continents except Antarctica.
Method: Provide wooden logs, boards, or blocks (typically within a house, barn, or shed), and allow the bees to bore in and create nests of their own.

Wool carder, leafcutter & mason & orchard bees
(Anthidium, Megachile, and *Osmia)*
Range: Anthidium and Megachile, all continents except Antarctica; Osmia, North America.
Method: Place hollow tubes ¼ inch (6 mm) in diameter, either singly or clustered, between knee and shoulder height above the ground. Drilled holes about 6 inches (165 mm) deep in wooden blocks may also work. Allow the bees to move in by themselves.

decisions to consider when setting up trap nests. For example, hanging hollow pieces of bamboo about 5 feet (165 cm) off the ground attracts wool carder bees (*Anthidium*). This height is important, because if they are placed any lower they will be within reach of predatory ants, and they will also tend to show a greater incidence of spider webs.

No matter what the hive design, it must have an entrance hole, at minimum. Added ventilation holes are beneficial in hot areas, and even in cold areas they are a good idea, to allow moisture to escape and thus to help to prevent fungal growth.

Ancient Hive Designs

Early beekeepers, for example in Egypt and Central America, used simple containers to shelter bees (typically honey bees and stingless bees) in nesting spaces reminiscent of natural tree hollows. Within these cavities, the worker bees built wax comb for brood and food storage. Materials for these hives varied by culture, but typically they were made from various combinations of hollow logs, sticks, bamboo, straw, stone, animal dung, mud, and earthenware. Simple designs like these are still used today, especially in developing countries.

FIXED-COMB DESIGNS: LOGS, SKEPS & BOX HIVES

By the Middle Ages, bees were being kept in log hives, which emulated the bees' natural home but were more convenient for the beekeeper. Skep hives also became a common sight in European apiaries. These hives, which resemble inverted baskets, were often made from woven wicker or coiled straw. Additional sections could be added to the base, or a top cap could be added, to enable expansion for brood or honey respectively. Skeps were sometimes given additional protection by being placed in wall recesses, known as bee boles. Other basic hive designs include various box hives such as the Johnson hive, still commonly used in Uganda. In all these hive designs, the bees make their own wax comb.

Hives such as these provide a natural and inexpensive improvement on laborious honey-hunting, but the fixed nature of the

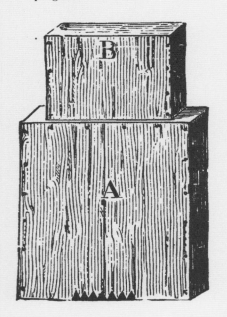

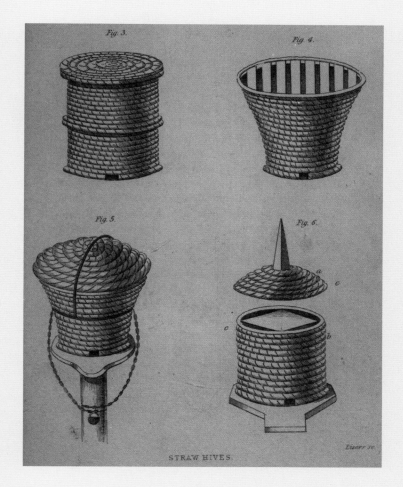

STRAW HIVES.

the seventeenth century, the octagonal Stewarton hive, designed in Scotland in the early nineteenth century, and the Warré or "people's hive" in the twentieth century. Additional boxes can be placed above or below the box containing the brood—known as supers or nadirs, respectively.

MOVABLE-COMB DESIGNS: TOP-BAR HIVES

Some ancient Greek hives, although made of wicker in the same way as skeps, used wooden top bars from which the bees would suspend their combs, allowing them to be removed for inspection. The sloping sides of the hives were designed to prevent the bees from attaching the combs to the sides. Modern versions of these, such as the "Kenyan" top-bar hives, are widely used in Africa. They have the advantage that they can be constructed easily from a vast range of materials, by beekeepers with little training. The only vital measurement in a hive like this is the spacing of the top bars.

Above *Straw hives were common in Europe from the Middle Ages to the nineteenth entury, and could be made to many different designs.*

combs requires destructive honey and wax harvesting, often killing bees or leaving them homeless. Additionally, these hives cannot be thoroughly inspected for disease, prompting some governments (such as in some parts of the United States) to enact laws prohibiting fixed-comb hives in favor of designs with movable combs that can be closely inspected for pests and diseases.

Later versions of box hives featured the innovation of stacked boxes, allowing honey to be harvested without disturbing the brood nest. These included a design by Sir Christopher Wren in

Right *A Kenyan top-bar hive.*

Langstroth Hives 🌿

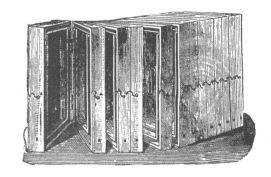

In 1789, Swiss naturalist François Huber revolutionized beehive design by introducing movable combs. The combs in his leaf hives were set within frames that opened up like the pages in a book. Huber designed his hives as an aid to studying the natural history of the bee, rather than for use in beekeeping, and they proved difficult to use—more of a curiosity than a practicality. Huber's writings, however, inspired the Reverend Lorenzo Lorraine Langstroth to come up with a new beehive design, and Langstroth hives are now widely recognized for making beekeeping a more viable and easily regulated practice.

BEE SPACE

In 1852, Langstroth patented the first beekeeping system to incorporate "bee space," a measurement of ⅜ of an inch (slightly less than 1 cm) that defines the minimum gap through which a bee can move. Honey bees add wax comb to spaces that are wider than ⅜ of an inch and fill in useless spaces smaller than ⅜ of an inch with propolis resin. By such means they maximize the opportunities for brood and nutrient storage, as well as limiting the space available to hive invaders. Langstroth was not the first to observe this behavior, but he was the first to turn it to practical use in beekeeping.

COMPONENTS

Because it incorporated proper "bee space," the wooden Langstroth hive design made honey extraction easier by ensuring that the frames were fully

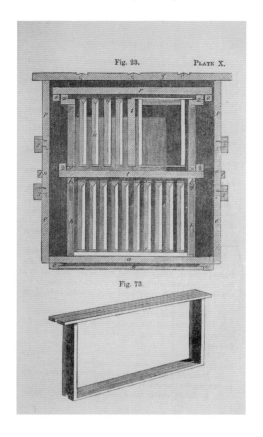

Fig. 23.   PLATE X.

Fig. 73.

Left *Fully removable frames are a key element of the Langstroth hive.*

removable. The design features two interior ledges, called runners, positioned opposite one another at the top of the hive box to support the top bars of the wooden frames, which hang freely with their wax comb. The other three sides of each frame are distanced a "bee space" from the hive walls to prevent comb or propolis deposition. Individual frames can thus be completely removed from the hive with ease. Modern frames typically have a plastic or wax foundation that is pre-molded with a uniform hexagonal pattern to give bees a base from which they can draw out comb. The hive box rests on a bottom board with a raised edge to create proper ventilation through a small hive entrance at the front.

EXPANSION

A feature of Langstroth hives is their expandability. When a colony is overcrowded, queen mandibular pheromone (QMP) does not diffuse through the hive thoroughly, and the colony may decide to make swarm queen cells to prepare for a hive split. Langstroth's design allows for several additional boxes, called supers, to be added to the top of the hive. Expanding the hive in this way prevents overcrowding and gives the colony additional room for brood and honey storage.

Above Hive design must take account of the fact that bees build comb bridges across spaces that are wider than "bee space" (⅜ inch/1 cm).

Right Langstroth hives can be easily expanded by adding additional supers on top of the brood box or boxes.

Observation Hives

Observation hives have been around since the seventeenth century. These transparent beehives were used as elaborate public displays. They were often fully fledged honey bee hives, with a glass window or glass globes through which the combs could be seen.

MODERN OBSERVATION HIVES

A modern observation hive consists of one or more frames in a box with glass on both of the larger sides. Sometimes the boxes are stacked on top of one another. With this system, it is possible to watch almost all the day-to-day activities of a colony of bees, in real time.

Until recently, observation hives have for the most part been set up as temporary structures. For research purposes, they are often kept only as long as an experiment lasts. As a publicity stunt, they are used for short stints at county fairs and other commercial exhibits. In museums, frames may be borrowed from a Langstroth hive and set up as observation hives, and subsequently returned to the larger site.

Now, by carefully monitoring and assisting thermoregulation (with insulation in the winter and ventilation in the summer), it is possible to keep an observation hive going for years. In addition, since they can be isolated into single, self-contained units, they can be made very safe for a classroom setting.

LEARNING OPPORTUNITIES

Learners of all ages can observe the honey bees' most intimate interactions and nesting life through the looking glass of an observation hive. Museums,

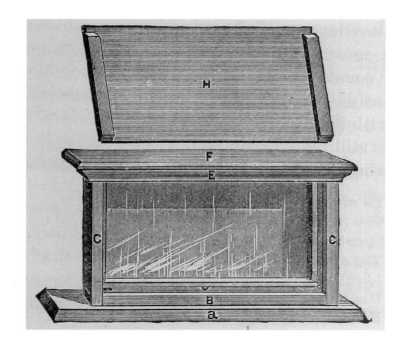

Below *An observation hive has glass on both sides to allow the activities of queen and workers to be closely watched.*

Above *A glass-encased bee colony being used as part of a demonstration of beekeeping at the Smithsonian Institution's Folklife Festival.*

classrooms, and public events are common places to find these structures. This is one of the few opportunities for viewing the inner workings of a "superorganism" (see Chapter 3).

For example, viewers can see how the work of the hive is performed—how food is distributed, how worker bees groom other bees, and how they clean the hive to prevent disease. Combined, these actions are analogous to the circulatory system in vertebrates—with a major difference being that it is all happening in full view.

Observation hives can be useful tools in education, covering a surprising range of topics. Take geometry, for example. Bees build hexagonal prisms to house their larvae and nectar. Why a hexagon, and not a triangle, square, or other shape? (The answer, in simple terms, is that packed hexagons not only provide the largest possible cells but also tessellate— i.e., fit together with no gaps.)

CLASSROOM OBSERVATION-HIVE CURRICULUM

In a math class, students would:

1. Calculate the surface areas of triangles, pentagons, hexagons, etc.

2. Use area formulas, volume formulas, and trigonometry to do the calculations.

3. Use the bee dance to learn about the use of angles in nature.

In an art class, students would:

4. Learn about Escher and the tessellations in his artwork.

5. Explore the use of bee products in art, such as encaustic.

In a writing or biology class, students would:

6. Describe the life of the honey bee, taking in queens, workers, and drones,.

7. Use red light to see the natural behavior of bees in the hive (honey bees cannot see red).

8. Experiment by changing variables such as temperature, light, noise, or vibration.

In a chemistry class, students would:

9. Study the chemistry of honey and pollen and what makes them such healthy products.

10. Create different types of liquid and solid feeds and record what the bees prefer or avoid.

Data Tracking 🐝

Record keeping is certainly not required, but it is recommended good practice. These records will help the beekeeper to understand the normal cycle of a beehive, as well as what to expect for the next visit. The predictive value of recording data is perhaps its greatest benefit. A knowledge of annual trends, and the difference between normal and abnormal years, allows the beekeeper to plan necessary interventions from year to year.

ECOLOGICAL TRADE-OFFS

An ecological approach to beekeeping relies on understanding the balance between the bees' investment in three critical aspects of life: growth, defense, and reproduction. Beekeepers should take care that their hives are developing in all three areas.

WHAT COULD GO WRONG?

The two specific things to look out for when checking a beehive are (1) disease, and (2) signs of swarming, both of which can develop between regular hive checks, so consistent diligence is necessary.

Left unchecked, hives can become diseased, and can then infect other hives for miles around, leading to dramatic

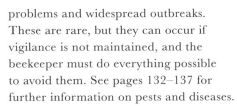

WHAT TO RECORD

As with all things beekeeping, personal style is important. Whether using paper and pencil or typing notes into an electronic device, record keeping helps beekeepers keep track of how their hives are doing. Core metrics focus on productivity and fitness. Additional measures of local ecology can help to predict annual trends, including floral blooms, temperatures, and hive disposition (aggressive, docile, etc.). Customizable mobile apps are available that automatically record date, time, latitude/longitude, and even weather and daylight conditions.

Core metrics

PRODUCTIVITY
- Honey
- Pollen

FITNESS
- Brood
- Adults
- Queen cells

HEALTH
- Subjective condition (e.g., excellent, good, troubled, dead)
- Disease observations

SWARM PREVENTION
- Number of queen cells present
- Size of colony

problems and widespread outbreaks. These are rare, but they can occur if vigilance is not maintained, and the beekeeper must do everything possible to avoid them. See pages 132–137 for further information on pests and diseases.

Swarming involves losing half the bees in the natal hive, when they leave to found a new colony (see pages 128–129). Both disease and swarming can directly impact the survival and productivity of the hive, leading to reduced honey yield and potentially dead hives. Check in with your hives about every ten days to prevent such problems.

Two other somewhat common maladies of honey bee colonies involve (1) losing the queen, and (2) very little honey production. Signs that a queen is weak or gone often involve a drastically increased abundance of drone brood, which is larger and more bulbous in appearance than worker brood. Worker bees are able to lay unfertilized eggs, but typically do not when queen mandibular pheromone (QMP) is present. In its absence, worker egg-laying occurs en masse, leading to the failure of the colony, since only drones are produced. Healthy workers are required in order for a hive to survive and produce honey. Other factors that could contribute to these problems are infection (*Nosema ceranae* is a likely culprit) and lack of suitable forage.

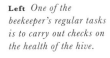

Left *One of the beekeeper's regular tasks is to carry out checks on the health of the hive.*

Urban Beekeeping

Paris is a famous hub and a global model for the practice of urban beekeeping. The rooftop apiary at the Opéra Garnier was established in the 1980s. In the years since, the government of France created a national program to promote urban beekeeping in all French cities. Meanwhile, on the other side of the Atlantic, beekeeping was illegal in New York City until March 2010.

What makes urban beekeeping unique is the public relations challenge. There is a distinct difference between the reality of bees and the public perception, and the common misconception that bees are dangerous is not just a recent phenomenon—many people have a deeply ingrained but quite unjustified fear of bees. This is troubling, especially given recent data showing that bees may survive better and make more honey in cities than in the countryside, a trend that has been noted on both sides of the Atlantic, from Paris to Boston.

SURVIVAL & PRODUCTIVITY

Between 2010 and 2013, urban beehives in Boston and Cambridge, Massachusetts, USA, survived the winter better than those in surrounding suburban and rural towns, at a rate of approximately

Above *An urban community garden can provide good bee habitat.*

beekeeping and to allay public fears. Guidelines tend to limit the number of hives permitted, such as one hive per lot in Cleveland, two in San Diego and Denver, four in Seattle. Another key condition often seen in urban beekeeping regulations is the creation of a flight path, whereby bees are directed to fly up and away from neighbors or walkways. This can easily be achieved by means of creative landscaping, using an obstacle about 1 foot (30 cm) away from the opening of a beehive.

two-thirds surviving in the city compared to only around two out of five in the countryside. Urban beehives in Boston and Cambridge also produce more honey than those in the surrounding areas, by about one-third more. Out of two hundred hives spread throughout eastern Massachusetts, the top honey-producing hives are consistently in either Boston or Cambridge. The survival and productivity rates vary from year to year, but the trends persist. Why is this? Some of the explanations underlying the success of urban bee populations are listed in the box on the right.

LEGAL ISSUES

The growing popularity of urban beekeeping has prompted governments around the world to create guidelines and recommendations for responsible practices. The legislation varies widely from country to country, and from city to city: in some places there is a full ban, while elsewhere beekeeping is completely unregulated, and there are all sorts of in-between situations. The goals of legislation are to provide recommendations for responsible

Left *Thriving beehives in an urban setting in Chicago, Illinois.*

URBAN ISLANDS

Hypotheses for why cities might be better for the survival and productivity of bees include:

- Warmer (urban heat-island effect).

- Less pesticide use than on agricultural lands.

- Better diet:

  Quantitative: more flowers.

  Qualitative: varied diet from community gardens (diverse pollen sources aid honey bee immune function).

- Less energy expended on foraging if flowers are closer to hives.

Harvesting 🐝

Some species of bee store food as a means to provision for future generations, as well as to cover times when pollen or nectar are in short supply, such as the winter, the rainy season, or during seasonal breaks in floral blooms. Honey bees in particular are tremendously efficient food storers, partly because there are tens of thousands of workers per hive, actively foraging and storing food. The beekeeper's challenge is to understand how much food the bees need to survive these dearth periods, and how much can be harvested for human consumption. As well as honey and pollen, beekeepers also harvest other bee products, including wax, propolis, and venom.

HONEY

Honey bees create wax combs made from hexagonal cells. Wax cells can be filled with brood, or with food (honey or pollen). Honey and wax can be harvested concurrently, while pollen is most easily collected separately.

To collect the honey, remove a frame or slat of honeycomb from the nest, shake or brush the bees off, and transport it to an area without bees, such as an indoor kitchen. The trouble with extracting honey outdoors is that it attracts bees, which can become a nuisance. However, outdoor messes tend to be cleaned up easily, with a little help from nearby bees.

To extract honey from wax comb, first scrape off the wax cappings from the tops of each cell that contain honey. Use a fork, a heated knife, or similar tool to do this. Large-scale beekeepers with many hives may invest in honey extraction

Left *Collecting the honey starts with removing the wax capping from the top of the comb.*

Right *Propolis, made
by honey bees from tree
resins, has antimicrobial
properties.*

equipment that can harvest hundreds of
pounds of honey at a time, but this is not
necessary for a smaller-scale enterprise.
When dealing with the output of just a
few hives, the honeycomb can be scooped
or pressed into a bowl, and then filtered
(cheese cloth or pantyhose work just fine).
Honey requires no heating, no additives,
no processing of any kind, and it is ready
for bottling or eating right away, fresh out
of the hive.

WAX

Wax accumulates as a by-product of
honey harvesting, and also from the comb
itself. Such wax may contain impurities
from debris, including pollen and dirt,
and this must be filtered out in order
to produce pure wax. Purifying wax is
simple. Melt solid wax in boiling water,
then pour the liquid mixture through a
filter or strainer into a clean pot. As the
liquid cools, the wax rises to the top of
the water and hardens into a pure brick.
Pure beeswax candles burn dripless and
smokeless. Wax can also be made into
lotions, balms, polishes, coatings, and
preservatives, among many other uses.

Right *Beekeeper Ian
Bailey watches as his
honey—made by bees
in London, UK—runs
through a sieve.*

POLLEN,
PROPOLIS & VENOM

Protein-rich pollen is most easily collected
using a trap device. The basic design of a
pollen trap involves getting the bees to walk
through a mesh and over a grated floor,
which knocks the pollen off their hindlegs
and into a collection drawer.

Propolis is a natural form of self-
medication for bees. Also known as "bee
glue," it is a sticky substance made from
tree resins that has antimicrobial properties.
It can be chipped off the inside of the
hive and made into gums, tinctures,
or creams to aid in wound healing.

Bee venom is of medical benefit to a
wide variety of human health concerns,
especially inflammatory and autoimmune
conditions. Symptoms of arthritis,
multiple sclerosis, amyotrophic lateral
sclerosis (motor neuron disease, Lou
Gehrig's disease), and even HIV can all
be treated with bee venom therapy. The
inflammatory reaction to venom has also
been used by the cosmetics industry as a
means of plumping lips and smoothing
wrinkles. Harvesting venom is technical
and difficult, involving killing worker
bees by using alarm cues to entice them
to sting a cloth over a collection jar.

"Natural" Beekeeping

Beekeeping today is different from the way it was in our grandparents' day. The near-global spread of pests and pathogens has brought new diseases and challenges. Recent advances in agricultural science have produced new treatment options for bee diseases—but if they are not used properly, these chemicals can have deleterious effects.

Chemical-free, organic, or "natural" beekeeping practices are a viable option for many beekeepers around the globe. Remember that honey bees have managed without human intervention for tens of thousands of years. However, the global spread of *Varroa* mites, and the pathogens they carry, has forced some beekeepers to advocate using in-hive chemical treatments and other interventions. Beekeepers choosing a strictly natural route, on the other hand, tend to justify their approach by referring to natural selection—let troubled hives go, and promote healthy hives by splitting them or rearing queens from them.

BASIC METHODS

Building up a strong and healthy hive is the ultimate goal of all beekeepers. Beekeepers should research their local area's disease trends, and inspect hives most frequently during periods of low population coinciding with cool and damp weather, when diseases are most likely to develop. Typically, catching a disease early can allow the simple solution of replacing any diseased parts of the hive. A less "natural" approach relies on applying synthetic chemicals inside the hive to clear the infection.

ARTIFICIAL SELECTION

Bee breeding programs might seem anathema to natural beekeepers, but those who believe in chemical-free practices appreciate the value of selecting for healthier genetic lines. There are at least three lines of bees in the USA that have been bred for improved resistance to parasites and diseases, especially the *Varroa* mite, American foulbrood, and chalkbrood. Find out more about the work of Dr. Marla Spivak and her team at the University of Minnesota, and the Minnesota hygienic bees, on page 211.

Beekeeping using only natural practices is sometimes known as chemical-free beekeeping. Going one step further, organic beekeepers do not add anything extra to a hive, beyond what is required for a complete housing structure. Standards for organic beekeeping practices vary globally. For example, if a hive is running low on stored food, then adding sugar water (supplemental carbohydrates) or pollen patties (supplemental proteins) is a chemical-free, but not an organic beekeeping method. Beekeepers who wish to adhere to full organic standards provide supplemental nutrition for their bees only in the form of frames of honey or bee bread (stored pollen within wax comb).

EVOLUTIONARY SELECTION

Darwinian natural selection provides a framework for understanding the natural beekeeping approach: let the strong hives reproduce, and allow the weak hives to die off. This can be a difficult mantra for beekeepers, who are devoted to doing anything they can for the health of their bees, including medicating and feeding them. However, the future of beekeeping relies on strong beehives, and that in turn comes down to genetics. Some hives have genes that are better suited for hygienic behavior (for example, detecting, uncapping, and removing diseased bees) and aseptic behavior (for example, grooming nest mates and cleaning out the nest). Selective breeding of beehives can help to pass these desirable genes on to subsequent generations, along with other helpful genes such as those that predispose to a gentle temperament and/or good honey production.

Above *Dr. Marla Spivak of the University of Minnesota carries out research on the artificial selection of disease-resistant bees.*

Swarming

When a beehive swarms, tens of thousands of bees depart the hive in a chaotic, aerial whirl, eventually settling on a nearby structure such as a tree branch or street sign for a few hours or a couple of days before finally relocating to their new nest site. Swarming is a natural part of honey bee biology—it is a means for the hive to reproduce—but beekeepers should make efforts to prevent swarming, or they risk a dramatic decline in hive population that can lead to drastically reduced honey yield, or even hive death. In urban environments, swarms can create problems owing to the perceived risk to the public. In reality, swarms are remarkably docile, and can be fun to collect.

Below *A swarm can be collected in a cardboard box, and then transported to a hive to settle in. Swarming bees are surprisingly non-aggressive.*

WORKING WITH SWARMS

Honey bee swarms can be caught, combined, or created.

Catching swarms

Before departing their natal hive, swarms gorge on honey to provide enough caloric fuel to last them through their journey to a new home, and also to help them produce wax to begin building the new comb. Swarms also have no hive to protect. The bees hang on one another, dangling from a structure waiting for scout bees to find a new hive space. Just as humans tend to reduce activity after very large meals, so do bees when swarming. Their relatively gentle nature allows for even novice beekeepers to brush, shake, or otherwise guide the bees into a box held just below the swarm. This box can then be covered and transported to an available hive, where the bees can be left to settle in and establish their new nest.

Below The creation of queen cells may be a sign that swarming is likely to occur.

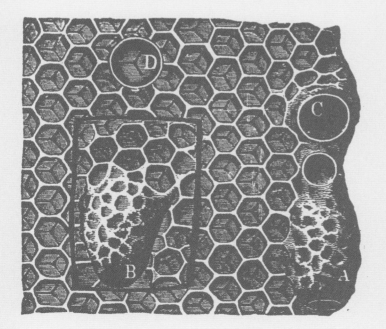

Combining swarms/hives (newspaper method)

In the rare event of two swarms being caught, or when two established hives are weak, they can be combined to make one stronger hive, using the newspaper method. A newspaper–honey combination is used as an edible barrier that delays direct interaction between two unrelated hives so that they can get used to each others' scents (and/or achieve the same scent, perhaps). Place newspaper over a hive with a nest of bees settled inside. Use sugar water or honey (from a healthy source hive!) to glue this thin piece of newspaper in place. Next, spray or coat the paper with more of the sweet substance. Lastly, place the second swarm, or weak hive, in a hive box over this. The bees will chew through the newspaper after the two groups have had a period to acclimate to one another, and fighting is not expected to ensue between them.

Preventing & creating swarms

When a beehive is growing rapidly in population, they will often create queen cells, which look like peanuts hanging down from the wax comb. A good management technique to prevent swarming—and losing half the bees—is to remove these queen cells. Diligence is key, as the hive will likely go on to make more queen cells.

However, if the hive is especially strong, then it might be desirable to split it into two or more hives by simply moving the frames with new developing queens into separate hive boxes. These new hives should be positioned as far as possible from the natal hive, at least temporarily, to help the foraging bees return to their new hive.

Integrated Management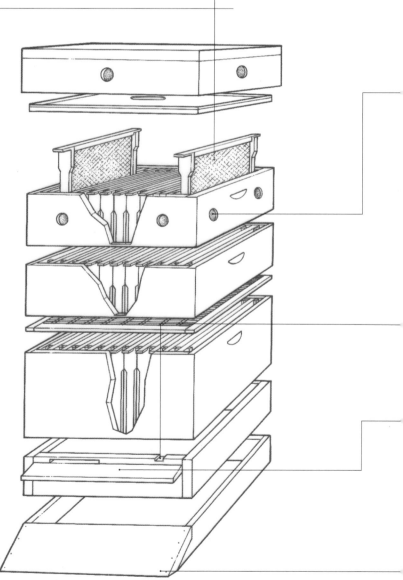

In the modern world, beekeeping has a whole new set of challenges to deal with. Beekeepers today are faced with more diseases, more agrochemicals (pesticides, fungicides, herbicides), and in many cases dramatically reduced habitat in which their bees can forage. The common goal among all beekeepers is for sustainable bee populations that are healthy and productive. Multiple challenges are best combated using multiple solutions. An integrated approach to rearing healthy bees is often the answer.

OVERWINTERING SUCCESS

In any part of the world, regardless of geography, "winter" from the perspective of the bees can be considered as any period when a shortage of forage is combined with inclement weather that forces them to stay in the hive. Whether it is excessively cold or there is too much rain, this time of year can be worrisome for beekeepers, as the bees have only their stored food and their nest mates to rely on for energy, warmth, dryness, and health. On top of that, insectivorous pests such as mice tend to try and get at the stored honey in the winter. That can lead to a hive's demise, as can blocked entrances and excessive moisture.

REPLACING EQUIPMENT

Wax can accumulate pesticides, so frames should be replaced every three to five years; pine and other woods rot over time, and hives therefore need to be kept in good repair.

MOISTURE CONTROL

Drilled holes or other ventilation at the top of the hive allows air to circulate and moisture to escape, minimizing the risk of microbial growth.

ENTRANCE REDUCER

A specially carved wooden entrance reducer, or a piece of metal with holes just large enough for the bees to pass through, can be placed in the hive opening to limit the area that guard bees are required to protect from intruders during the winter. Keep the hive entrance clear of snow and debris so that bees can come and go—they make their "cleansing flights" year-round to defecate, so as to keep the inner hive aseptic.

DEBRIS DISPOSAL

Sticky paper and a screened bottom board may be used to prevent debris accumulation, and to monitor for Varroa *mites knocked off during bee grooming.*

PROTECTING FROM THE ENVIRONMENT

Beekeepers in cold regions often wrap the outside of their beehives with insulation to provide additional warmth during winter. Beekeepers in wet tropical regions may elevate their hives above the flood level, and hives are also suspended at a height to keep them away from ground predators—as seen in the hanging log hives of Kenya.

SELECTING THE SUBSPECIES

Many different types of honey bees are recognized as distinct races, or subspecies. These originate in different parts of the world, and vary slightly with regard to the environmental conditions they thrive best in, as well as in other behavioral traits such as temperament, swarming, production of honey/wax/propolis, and disease resistance.

EUROPE
Apis mellifera ligustica (Italian honey bee)
Apis mellifera carnica (Carniolan honey bee)
Apis mellifera caucasica (Caucasian honey bee)
Apis mellifera mellifera (Dark European honey bee)
Apis mellifera remipes (Persian honey bee)
Apis mellifera iberiensis (Spanish honey bee)
Apis mellifera cecropia (Greek honey bee)
Apis mellifera cypria (Cypriot honey bee)
Apis mellifera ruttneri (Maltese honey bee)
Apis mellifera sicula (Sicilian honey bee)

AFRICA
Apis mellifera scutellata (African honey bee)
Apis mellifera capensis (Cape honey bee)
Apis mellifera monticola (Mountain honey bee)
Apis mellifera sahariensis, Apis mellifera major (Moroccan honey bee)
Apis mellifera intermissa (North African honey bee)
Apis mellifera adansonii (Nigerian honey bee)
Apis mellifera unicolor (Madagascar honey bee)
Apis mellifera lamarckii (Lamarck's honey bee)
Apis mellifera litorea (East African lowland honey bee)
Apis mellifera nubica (Nubian honey bee – Sudan)
Apis mellifera jemenitica (Central African honey bee)

MIDDLE EAST & ASIA
Apis mellifera macedonica (Macedonian honey bee)
Apis mellifera meda (Iraqi honey bee)
Apis mellifera adamii (Cretan honey bee)
Apis mellifera armeniaca (Armenian honey bee)
Apis mellifera anatolica (Turkish honey bee)
Apis mellifera syriaca (Syrian honey bee)
Apis mellifera pomonella (Far East honey bee)

Arthropod Pests

Honey bees have been thriving without humans for much longer than with us, so they are well able to stay healthy. In modern beekeeping, the emphasis needs to be on guarding against a small number of relatively new pests to beehives, including wax moths, hive beetles, and parasitic mites.

WAX MOTHS

The wax moths *Achroia grisella* and *Galleria mellonella* develop from white caterpillar larvae, which grow to a length of an inch (25 mm) or so. The larvae eat their way through wax frames, leaving behind destructive trails of webbing, feces, and shredded comb. The adult moths do not cause harm directly. Worm-like larvae, cocooned pupae, and brownish adults are all easily spotted, and should be manually squashed on sight.

Right *The remnants of honey combs destroyed by wax moths.*

Below *The developmental stages of a wax moth, from egg to larva to adult.*

HIVE BEETLES

Both the small and large hive beetle species are native to Africa. The small hive beetle (*Aethina tumida*) spread to North America in the 1990s and to Australia in the 2000s. It is approximately ¼ inch (6 mm) long but can significantly damage frame integrity, brood viability, and food storage purity (honey and pollen) as it tunnels through cells. Its grub larvae resemble those of wax moths but are

Below *Small hive beetle* (Aethina tumida).

half the size and are usually found in the corners of frames. The large hive beetle (*Oplostomus* spp.) is approximately ¾ inch (20 mm) long, and is currently limited in range to Africa.

Beekeepers should remove beetles and larvae, or simply squash them in place and leave the mess for the bees to clean up. Beetle traps and organophosphate chemical treatments are also available, but a hive beetle infestation may be difficult to eradicate.

VARROA MITE

Varroa destructor mites are a critical and worldwide beekeeping concern. They are native to southeastern Asia, but have spread throughout the world since the mid-twentieth century. Adult female mites are reddish-brown, oval, the size of a pinhead, and often seen on the dorsal thorax or abdomen of worker bees throughout the hive. Females enter brood cells prior to capping and make a permanent hole in the developing pupa's exoskeleton to take blood meals for egg-laying nourishment. These wounds make the bee weak, deformed, and susceptible to infection by myriad pathogens carried by the mites, especially deformed wing virus. *Varroa* eggs will hatch, feed on the developing bee, grow, copulate with their sibling mites, and then exit the cell on the back of the young bee as it emerges.

Available treatments for *Varroa* mites include dusting with powdered sugar (to increase bee grooming), vaporized oils (e.g., thymol), acetic acid, sucrose octanoate, screened or sticky bottom boards, drone frames, pyrethroids (e.g., fluvalinate), oxalic acid, and organophosphates (e.g., coumaphos)— though not all of these have been proven to be effective. Building up strong hives is a good way to counter *Varroa* infection, allowing the bees to handle it themselves, when possible.

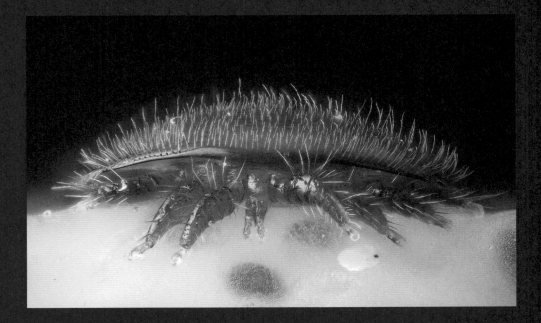

Right *Adult female* Varroa destructor, *head-on view.*

Bacterial, Fungal & Viral Infections

Many different pathogens infect bees around the world. Most infections are self-limiting, meaning bees can recover on their own. But some infections require intervention by the beekeeper.

AMERICAN FOULBROOD

The most infectious bee disease is American foulbrood (AFB). The responsible bacterium, *Paenibacillus larvae*, forms persistent spores that can remain infective for up to sixty years, making this disease highly contagious. Signs of infection include a very spotty brood pattern, dark and sunken brood cappings, and decomposed brown jelly-like brood. In many countries, legislation requires the burning of AFB-infected hives to prevent the spread of the disease.

Antibiotics (oxytetracycline) are permitted in some parts of the world, such as in the USA, but could weaken the hive against reinfection and contaminate honey if misused. An easy test for AFB is to poke a suspected area of brood with a matchstick, with a positive result if a string of goo appears when the stick is removed. Alternatively, a modern test kit is now available and reliable.

EUROPEAN FOULBROOD

The bacterium *Melissococcus plutonius* causes European foulbrood (EFB). EFB-infected larvae become discolored and misshapen, and die before capping. *M. plutonius* is a gut parasite that interferes with food uptake, and it can cause death if food availability is limited. A spotty

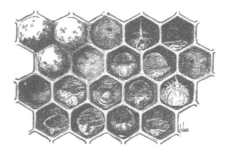

Above *American foulbrood, caused by* Paenibacillus larvae.

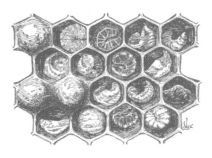

Above *European foulbrood, caused by* Melissococcus plutonius.

Above *An adult honey bee (*Apis mellifera*) showing the crumpled wings characteristic of deformed wing virus infection.*

brood pattern could indicate EFB (although it can also be a sign of inbred stock). Strong colonies should be able to overcome EFB on their own, but beekeepers can assist the hive by removing any infected comb manually. Antibiotics (oxytetracycline) are legal only in some parts of the world to treat EFB, but should be used responsibly. One method that has been effective at clearing EFB is the "shook swarm" or "shakedown" technique, whereby the beekeeper shakes bees off old comb and onto new comb in order to replace old and seasoned beehive parts clean new ones.

CHALKBROOD

Chalkbrood is caused by the fungus *Ascosphaera apis*, which infects the digestive system of bee larvae and causes malnourishment before ultimately consuming the bee itself. In addition to white fungus left in brood cells, mummified dead bodies are visible either within the comb cells or, most often, removed and carried to the front of the hive by workers. There is no common treatment for chalkbrood, so removing infected frames and increasing ventilation is the best course of action to help weakened colonies. As with other brood infections, replacing the queen can clear the infection as well. Moving the hive to a dryer and warmer location may also help to eliminate the fungus.

NOSEMA

Nosema apis and *N. ceranae* are fungi that cause serious digestive infections. Infected bees become malnourished and less productive, and have shorter lifespans, although they typically display few or no other symptoms. *N. apis* infection is often associated with dysentery in the colony: yellow, orange, or brown streaks of pollen waste can be seen spread around the inside and outside of the hive. *N. ceranae* infection often does not show this symptom. The antimicrobial agent fumagillin can be used to limit fungal spore growth in countries where it is legal.

PARALYSIS & DEFORMED WING VIRUSES

Viral infections of honey bees are diverse and often difficult to detect. Of them all, however, the paralysis viruses and deformed wing virus are probably the easiest infections for beekeepers to spot. As the name suggests, paralysis virus infections, in general, result in bees behaving extremely slowly, looking like zombies, walking at a crawling pace, and not flying. This sometimes affects the entire colony and results in the loss of all bees. Chronic paralysis virus is thought to be less closely related to the other paralysis viruses, and is more likely to be transmitted from bee to bee. Deformed wing virus can be very common; typically, it does not affect all the bees in the colony, but rather just a few adults will show crumpled wings. These viruses and others are transmitted by *Varroa* mites, and so the best preventative measure against these infections is to limit the prevalence of the mites.

HONEY BEE DISEASE SYMPTOMS & INFECTIONS

Observation	Disease name	Pathogen type
Red/brown mite present	*Varroa* infestation	Arthropod
Black beetle present	Hive beetle	Arthropod
Tunnels bored through wax comb	Wax moth infestation (could also be beetles)	Arthropod
Fecal matter at the front of the hive	*Nosema apis* or fermented stores	Fungal
Brood discoloration: white, black, or chalky	Chalkbrood	Fungal
Brood discoloration: gray, stone-like	Stonebrood	Fungal
Brood discoloration (capped): greasy appearance, sagging	American foulbrood	Bacterial
Brood discoloration (uncapped): brown, twisted, sagging	European foulbrood	Bacterial
Brood discoloration: pearly sheen with a sac around larvae; "Chinese slipper" appearance of dried larvae	Sacbrood virus	Viral
Crumpled wings, stumpy body	Deformed wing virus	Viral
Hairless, black bees, dislocated wings, or zombie-like behavior	Chronic paralysis virus	Viral

NATURAL BEEKEEPING METHODS IN RESPONSE TO COMMON PESTS & INFECTIONS

Disease	Next visit scheduling	Details	Treatment
Small hive beetles	Normal 1 month scheduling unless dozens of larva present, in which case 2 weeks	It can potentially devastate colonies in certain locations. Make sure to squash all of adults and larvae	Fill shallow troughs with mineral oil and place in the hive
Varroa mites	3 weeks	These are in many hives, so note if they are on most bees	A range of methods including oxalic acid (see page 133)
Chalkbrood	3 weeks	This is typical during cold months. Should clear on its own if the hive is otherwise healthy	Requeen the hive with a queen of a different strain
American foulbrood	1 day	Take a matchstick and stick it into the rotten brood. Positive test if stringy goo comes out after pulling the stick out	Destroy the colony. Burn frames and thoroughly scorch other wooden hive parts
European foulbrood	1 day	Often endemic. Symptoms are masked in times of plenty and show up in times of food shortage	In severe cases treat as for AFB. In mild cases use the "shook swarm technique"
Stonebrood	3 weeks	Should clear on its own if the hive is otherwise healthy	Remove/replace infected frames
Deformed wings	3 weeks	Viral infection transmitted by *Varroa* mites	See treatment for *Varroa*
Paralysis	1 day	Zombie-like bees	Dump all bees out, equipment can be reused. Sometimes this infection can disappear spontaneously
Wax moths	Normal 1 month scheduling unless dozens of larva present, in which case 2 weeks	More of a nuisance than a threat	Squash all adults and larvae, replace equipment if necessary

HONEY BEE DISEASE TREATMENT OPTIONS: CHEMICAL VS. CHEMICAL-FREE, PREVENTATIVE VS. RESTORATIVE

For any given disease, the various approaches shown here are not mutually exclusive, and can be used together as a form of integrated pest management. Many beekeepers will use a combination of these biotechnical and chemical interventions based on their assessment of the severity of the pest or disease problem.

Disease	Preventative treatment (chemical)	Preventative treatment (chemical-free)	Restorative treatment (chemical)	Restorative treatment (chemical-free)
American foulbrood (*Paenibacillus larvae*)	None (although oxytetracycline has been used, overuse of antibiotics is not recommended)	Build up a strong colony, replace queen periodically	None	Burn or irrdiate the hive
European foulbrood (*Melissococus plutonius*)	None (although oxytetracycline has been used, overuse of antibiotics is not recommended)	Build up a strong colony, replace queen periodically	None (although oxytetracycline has been used, overuse of antibiotics is not recommended)	Burn or irrdiate the hive
Nosema (*Nosema* spp.)	Fumagillin (where legal)	Build up a strong hive, replace queen periodically	Fumagillin (where legal)	Build up a strong hive, replace queen periodically
Chalkbrood (*Ascophaera apis*)	None	Build up a strong hive, replace queen periodically	None	(1) Remove infected frames, (2) requeen the hive to break the brood cycle
Sacbrood (sacbrood virus)	None	Build up a strong hive, replace queen periodically	None	(1) Remove infected frames, (2) requeen the hive to break the brood cycle
Stonebrood (*Aspergillus* spp.)	None	Build up a strong hive, replace queen periodically	None	(1) Remove infected frames, (2) requeen the hive to break the brood cycle
Slow paralysis virus	None	Build up a strong hive, replace queen periodically	None	Replace bees
Black queen cell virus	None	Build up a strong hive, replace queen periodically	None	Replace queen
Varroa mite (*Varroa* spp.)	Apistan strips, Checkmite+ strips, Sucrocide, Oxalic acid, Formic acid, Vaporized mineral oil, Ascetic acid	(1) Build up a strong hive, (2) replace queen periodically, (3) douse bees with powdered sugar to promote grooming activity, (4) use a screened bottom board, (5) use a drone frame, (6) use small cell foundation, (7) requeen with a *Varroa*-sensitive hygienic genetic breed	Fluvalinate (Apistan strips), coumaphos (Checkmite+ strips), sucrose octanoate esters (Sucrocide), thymol (Apiguard), oxalic acid (Mite-away), formic acid (Mite-away), vaporized mineral oil, ascetic acid	(1) Douse bees with powdered sugar to promote grooming activity, (2) use a screened bottom board, (3) use a drone frame, (4) use small cell foundation, (5) requeen with a *Varroa*-sensitive hygienic genetic breed
Tracheal mite (*Acarapis woodi*)	Menthol crystals	Build up a strong hive, replace queen periodically	Menthol crystals	(1) Requeen with resistant genetic line, (2) grease patties (1 part vegetable shortening + 3 parts powdered sugar) prevent transfer of mites between bees
Small hive beetle (*Aethina tumida*)	None	Build up a strong hive, replace queen periodically	Coumaphos (Checkmite+ strips)	Oil traps
Wax moth (*Galleria mellonella*)	None	Build up a strong hive, replace queen periodically	Paradichlorobenzene (PDB) crystals (urinal cakes where legal and only in stored equipment and not in living colonies)	Remove and freeze infected frames

CHAPTER 6

A Directory of Bees

Solitary Bees

There are about twenty thousand species of bees, of which approximately 250 are bumble bees, 500–600 are stingless bees, and 7 are honey bees. The remainder— the biggest group by far—are the solitary bees. This is a very variable collection of species, not least in terms of physical size. The world's smallest bee is *Quasihesma* sp., only 0.07 inches (1.8 mm) long, found on gum trees in Queensland, Australia. In contrast, the female of the largest bee, Wallace's giant bee (*Megachile pluto*), is 1.54 inches (39 mm) long, with a wingspan of 2.48 inches (63 mm).

Some bees show stunning adaptations to suit their ecological niche, such as the abnormally large jaws that Wallace's giant bee uses to collect and manipulate resin. Some, such as *Colletes brevicornis*, are highly selective, needing a specific host plant to survive, while others are generalists and far more flexible in their food requirements.

We know surprisingly little about most species of solitary bee. In Latin America and Australia, many species have yet to be identified—and sometimes, sadly, a new species is identified only to be immediately declared extinct. The solitary bees that we know most about are those that impact on humans in some way, or those that are easy to find.

Above *Leafcutter bees are so named because they cut pieces of leaves that they use to line cells for their brood to live in.*

ECONOMIC IMPORTANCE

Given the wide range of solitary bees, comparatively few have been exploited by humans as pollinators—even though many are more efficient at pollination than the ubiquitous western honey bee (*Apis mellifera*). But some species do have economic importance. For example, the alfalfa or lucerne leafcutter bee (*Megachile rotundata*) is used to pollinate alfalfa in Alberta, Saskatchewan, and Manitoba, and the bees are produced commercially. They readily take to artificial nests, and the cocoons are processed in large factories. Other solitary bees, such as the blue orchard bee (*Osmia lignaria*), have recently been brought into commercial use, while *O. bicornis* has the potential to be exploited as an orchard pollinator.

ASPECTS OF SOCIALITY

Solitary bees vary greatly in many aspects of their ecology, including where they sit along an evolutionary line from completely solitary to fully social. The group may be termed "solitary," but some of these bees are in fact potentially social. Bees can be grouped into:

- **Truly solitary**: a single mated female raises a brood in her solitary nest.
- **Communal**: many mated females, each with her own nest, build their nests together in an aggregation.
- **Primitively social**: a mated female (a foundress) makes her nest, forages, and produces worker daughters, who then help her to rear subsequent broods.
- **Social**: a mated female (a queen) exists solely to produce eggs, supported by her sterile daughter workers for nest building, foraging, and tending the young.

PROTECTING THE NEXT GENERATION

Many solitary bees in temperate regions overwinter within their nest cells as prepupae, pupae, or adults. They go into a form of stasis called diapause, which ends when certain conditions are met. The next generation of bees can then emerge and mate, and a new cycle begins. The nests of solitary bees vary considerably from one species to another. They might nest in soil, in sand, in holes, in trees, or in rocky crevices. The depth and shape of the nest, the position and number of egg cells, and the way the cells are sealed also differ greatly. Many females line the nest to protect their developing brood through the winter. The leafcutter bees line their nests with pieces of leaf, while the plasterer bees use a natural polyester to waterproof the cell.

Body lengths have been obtained from a range of sources and many species display a great deal of variability. Many measurements have been made using specimens held at the Natural History Museum, London, and the American Museum of Natural History, New York. The lengths are the average total of the head, thorax, and abdomen.

(Female shown actual size)

Bryony Bee *Andrena florea*

BODY LENGTH
Female 0.51 in. (13 mm)
Male 0.43 in. (11 mm)

FOOD PLANTS A specialist, collecting pollen only from white bryony (*Bryonia dioica*) and occasionally *Bryonia alba*. Collects nectar from a wider range of plants.

HABITAT & STATUS Widespread but scarce in habitats with sandy or light loamy soils such as heathland, parks, hedgerows, woodland margins, and roadside verges.

Description A black bee with golden body fur and reddish bands at the top of the abdomen.

Behavior & life cycle The family Andrenidae is the largest group of solitary mining bees. More than 500 species are found in North America and some 360 species in Europe and Asia. Typically, they mate soon after emergence. The females can be seen digging the soil, using their mandibles to loosen the soil and their front legs to scrape it backwards. As the burrow becomes longer, the female climbs into the nest and uses her abdomen and back legs to move the loosened soil out of the burrow. She smooths and lines the cell walls using her trowel-like pygidial plate at the end of her abdomen and the terpenoids that she secretes. This lining protects her young from fungal attack and water. She provisions the cell with pollen and watery honey, lays an egg, closes the chamber, and then makes the next cell. Nest burrows are usually no deeper than 2 feet (60 cm). Adults of the bryony bee are active from May to August. The mated female mines a nest burrow in sandy soil. She produces a single brood in May/June. Nest burrows may be made singly, or in groups of up to 100 nests. They are inactive over the winter, with the developing bees held in a state of diapause until the spring. In Germany, nest burrows host the cuckoo bee *Nomada succincta*.

(Female shown actual size)

Tawny Mining Bee *Andrena fulva*

BODY LENGTH
Female 0.39–0.55 in. (10–14 mm)
Male 0.39–0.55 in. (10–14 mm)

FOOD PLANTS A wide range of plants including weeds, garden plants, shrubs, and trees.

HABITAT & STATUS Widespread and common in parks and gardens, including lawns, flower beds, and mown banks; also chalk downland, grassland, orchards, and the edges of sparsely cropped agricultural land.

Description The female has long, dense, bright reddish hairs on the back of the thorax and abdomen. The rest of her body and head is covered with black hair. The male is slimmer and dull brown, with sparse reddish-brown hairs and a tuft of white hairs on the lower face.

Behavior & life cycle This is another member of the Andrenidae family. A single brood is produced between late March and mid-June. The nests are placed in large aggregations. Bees fly out in the spring, from early April until early June, when apple, pear, and cherry trees are in flower. The males emerge first. Flying in zigzag patterns, they seek newly emerged females. The males can mate repeatedly but the females only mate once. The mated female excavates up to three nest burrows in level soil, leaving a mound of soil at each nest entrance, with a main shaft dividing into four or five side tubes. She waterproofs the walls with terpenoid secretions smoothed by her pygidial plate. Into each tube a single egg is laid. The brood develop into adults and overwinter in their cells. This species may host the cuckoo bees *Nomada signata* and *N. panzer*, bee flies (*Bombylius major*), and the anthomyiid fly (*Leucophora obtusa*). Female bees may also be infested with mermithid nematodes. The tawny mining bee does not sting.

(Female shown actual size)

Perdita minima

BODY LENGTH
Female 0.08 in. (2 mm)
Male 0.08 in. (2 mm)

FOOD PLANTS Collects pollen and nectar from a range of Euphorbiaceae, such as *Chamaesyce albomarginata, C. polycarpa,* and other euphorbia species. An occasional visitor to Asteraceae and Polygonaceae.

HABITAT & STATUS Locally common in sandy desert.

Description The female is golden-brown, and the male is golden-yellow, with sparse hair. May often be detected by its shadow as it flies over the ground.

Behavior & life cycle A mining bee in the Andrenidae family (tribe Panurgini), *Perdita minima* is so small that the pollen grains it collects are large in comparison to the bee, but the female succeeds in packing the hairy scopa on her hind legs with substantial pollen loads in order to provision the nest. Many Panurgini species are found in Europe and Asia, but the genus *Perdita* is unique to North America. The adults are normally only active for one month. *Perdita* bees burrow holes on vertical, sloping, or horizontal sites in sand. Some nests have a side entrance. The nests can be communal, with many individual nests in large aggregations, or alone. The female smooths the nest walls but does not line the nest cell. Instead she coats a pollen ball, which she shapes to surround the elongated white egg. She seals the nest with a spiral of earth and sand. The larva hatches and eats the pollen. The growing larva rolls onto its back and cradles the remaining pollen on its middle. This prevents fungal contamination of the uneaten food. Most Panurgini produce one brood per year, but some *Perdita* species produce two broods.

Female shown actual size)

Teddy Bear Bee *Amegilla bombiformis*

BODY LENGTH
Female 0.28–0.71 in. (7–18 mm)
Male 0.28–0.71 in. (7–18 mm)

FOOD PLANTS *Buddleia*, *Dianella caerulea* (blue flax), *Abelia* flowers, *Hibbertia dentate*, and *Senna clavigera*.

HABITAT & STATUS Common in parks and gardens, riversides, woodlands, forests, and heath areas.

Description A fast-flying bee that hovers, pausing to buzz on a flower only momentarily. It is plump and round-bodied with long golden-brown fur, and is sometimes called the golden-haired mortar bee. Darker bands bisect the areas of fur on the body, seven on the male and six on the female. Young bees are densely furred, but the fuzzy texture becomes worn on older bees, revealing more areas of black and diminishing the characteristic "teddy bear" effect.

Behavior & life cycle *Amegilla bombiformis* is a member of the Apidae family (tribe Anthophorini) that lives and reproduces in individual nests, often in shallow scrapes in the soil. It typically selects corners with some shelter or camouflage (in populated areas it may be found burrowing under the foundations of houses), or it may choose to nest in abandoned mammal burrows, or along the banks of creeks. Groups of females will often nest close by one another. The male does not use the nest but roosts on plant stems. The female excavates a tunnel in her chosen burrow and lays her eggs in secure waterproof oval cells, leaving a store of gathered flower nectar and pollen in each cell as supplies for the developing larva, before sealing it. They are not aggressive bees, and will sting only if caught or stood upon. This species is host to the domino cuckoo bee (*Thyreus lugubris*) (see page 152).

(Male shown actual size)

Blue-Banded Bee *Amegilla cingulata*

BODY LENGTH
Female 0.39–0.59 in. (10–15 mm)
Male 0.39–0.59 in. (10–15 mm)

FOOD PLANTS Tomato, kiwifruit, eggplant (aubergine), several Australian native plants, and a range of garden plants such as buddleia (*Buddleja* spp.) and lavender (*Lavendula* spp.). Prefers blue flowers.

HABITAT & STATUS Common in tropical and subtropical regions, found in urban areas, woodlands, forests, and heathland.

Description A fast-flying bee that darts and hovers between plants, *Amegilla cingulata* has striking iridescent blue stripes of fur on its black abdomen. It has a plump, round, furry body with red-brown fur on its thorax.

Behavior & life cycle *A. cingulata* is a member of the Apidae family (tribe Anthophorini). It nests in individual burrows formed in soil or soft sandstone, in muddy riverbanks, mud bricks, or soft mortar, sometimes under house foundations. Females typically build nests in a group. Each female excavates an individual tunnel and lays her eggs in secure waterproof cells, provisioning them with nectar and pollen to feed the developing larva which overwinter as immature bees. The male roosts on thin plant stems such as grasses. This bee normally flies only in the warm months (October to April), when they are important buzz-pollinators of crops, garden flowers, and native plants—which in Australia include guinea flower (*Hibbertia*), cassia (*Senna*), mountain devil (*Lambertia formosa*), and gray spider flower (*Grevillea buxifolia*). They usually die when it becomes cold, but in some areas adult bees remain active throughout winter. *A. cingulata* can sting, but they are not considered aggressive. This bee's nest is parasitized by the neon cuckoo bee (*Thyreus nitidulus*).

(Male shown actual size)

Hairy-Footed Flower Bee *Anthophora plumipes*

BODY LENGTH
Female 0.51–0.59 in. (13–15 mm)
Male 0.51–0.59 in. (13–15 mm)

FOOD PLANTS Many spring flowers, including *Primula*, lungworts (*Pulmonaria* spp.), borage (*Borago officinalis*), deadnettles (*Lamium* spp.), broad bean (*Vicia faba*), comfrey (*Symphytum* spp.), and rosemary (*Rosmarinus officinalis*).

HABITAT & STATUS Common and widespread in many habitats, including gardens, open woodland, and coastal sites.

Description Densely furry, and both sexes are distinctive. In the UK, the males are usually orange-brown, but elsewhere they are gray or black. The lower part of the male's face is yellow, and he has distinctive long black hairs on his elongated mid legs. Females are either black or brown with yellow hairs on their hind legs. The wide range of colors has led to this bee being classified as several different species—but it is now recognized as a single species, in the family Apidae (tribe Anthophorini). It is also known as the plume-legged bee.

Behavior & life cycle These are fast-flying solitary bees. Males hibernate over winter to emerge in late February or March, with females emerging a few weeks later to mate and nest. The males are highly territorial and will chase off any intruders from their area, which will include forage and a suitable nest site. The females form long nest holes in clay banks, mud walls, or soft mortar. The nest is subdivided into a sequence of cells, each containing a pollen and nectar mass on which is laid a single egg. The life cycle is completed by the end of June. The cuckoo bee *Melecta albifrons* will lay its eggs in the nest. Its larvae emerge faster than the host bee's larvae. The cuckoo bee larva eats the larval stores, pupates, and emerges the next year in place of the host bee.

(Female shown actual size)

Southeastern Blueberry Bee *Habropoda laboriosa*

BODY LENGTH
Female 0.59–0.63 in. (15–16 mm)
Male 0.51–0.55 in. (13–14 mm)

FOOD PLANTS Blueberry (*Vaccinium*). Will also visit other unrelated species for pollen and nectar, including *Gelsemium, Quercus, Cercis, Azalea, Cirsium, Lupinus, Malus, Melilotus, Prunus,* and *Vicia.*

HABITAT & STATUS Widespread in urban, suburban, and rural areas.

Description The female is black with pale fur on her thorax and black fur on her head, underside of thorax, and legs. The male is similar but has white fur on his face.

Behavior & life cycle This member of the family Apidae (tribe Anthophorini) is active from late February until late May. In Florida, they are seen from November to February. Their flight period matches the three- to five-week flowering period of the blueberry (*Vaccinium* spp.), which they buzz-pollinate. A single brood is produced each year. The males emerge up to eighteen days before the females. They fly in a zigzag pattern above the nest sites and form a mating clump around a virgin female on the ground. The mated female scrambles out to excavate a multi-chambered nest. The main shaft slopes down to two oval nest cells. They are lined with a thin protective waxy coating which permeates the cell walls. One-third of each cell is filled with a pasty pollen mass, upon which a single egg is laid, and the cell is capped with a waxy coating. The nest is left unplugged. By August, the nests contain prepupae. It is not known which developmental stage overwinters. Burrows may be single or grouped under forest litter, in open sand, or in earthen holes. Nests may be abandoned if they are raided by Argentine ants (*Iridomyrmex humilis*).

(Female shown actual size)

Eastern Cucurbit Bee *Peponapis pruinosa*

BODY LENGTH
Female 0.47–0.55 in. (12–14 mm)
Male 0.43–0.51 in. (11–13 mm)

FOOD PLANTS Requires *Cucurbita* species for pollen, preferring native squashes, then summer squashes, winter squashes, and pumpkins. Also visits *Pontederia, Asclepias, Blephilia, Verbena, Cephalanthus, Convolvulus,* and *Ipomoea.*

HABITAT & STATUS Widespread and common wherever squashes, pumpkins, and gourds are grown.

Description The female is black with pale golden-yellow fur, including the long scopal hairs. The fur is short, dense, and red-gold on the thorax with broad pale gold bands on the abdomen. The male has a yellow face and white hairs on the abdomen, and very long antennae.

Behavior & life cycle This species is classified in family Apidae, tribe Eucerini (long-horned bees). Adults emerge from June to September/October, when cucurbits are in flower. For the first two to three weeks after emergence the females sleep in the flowers, foraging from half an hour before sunrise until noon, when the flowers are open. The males fly around the flowers searching for a mate, and mating typically takes place in the flower. The males rest in the closed flowers from noon. The females rest away from the flowers but return to gather nectar at dusk. The mated females nest in aggregations, near or under cucurbit plants. The same sites are used for years. They may make more than one nest burrow, with an entrance turret, a vertical shaft, and four or five nest cells set at downward angles at the bottom. The female lays an egg on a pollen mass and plugs the nest. The young bees overwinter as prepupae and emerge the following summer. There is one brood each year.

(Female shown actual size)

Squash Bee *Xenoglossa fulva*

BODY LENGTH
Female 0.55–0.71 in. (14–18 mm)
Male 0.55–0.63 in. (14–16 mm)

FOOD PLANTS *Curcubita* spp. for pollen. Also seen on *Ipomoea, Asclepias, Salvia, Mandevilla,* and *Helmia foliosa*.

HABITAT RANGE Locally common and widespread, anywhere that curcurbits are grown.

Description These bees have unusually large ocelli, typical of bees that fly when light levels are poor. They are black with golden red-brown legs and dark feet. They have long, dense pale golden fur and yellow-brown wings. They are members of family Apidae (tribe Eucerini), but—unlike *Peponapis* species—the male has a black face and the female has a yellow face.

Behavior & life cycle Squash flowers open before dawn and wither until closed by noon. These bees are active when the flowers are open, foraging and mating before noon. They often shelter in the closed flowers, and must cut or tear their way out of the dead bloom. The female nests in loam soil in open grassland amid an aggregation of burrows from earlier years. She digs a vertical shaft, with a mound of excavated soil up to 4 inches (10 cm) tall around the entrance. Oval nest cells are formed off the main shaft. Each cell is lined with a protective substance and one-quarter filled with pollen mass and a thin layer of nectar. A single egg is then laid on top. The cells are capped and the burrow is plugged. The developing bees overwinter as larvae, transforming into adults during June/July, shortly before emergence in August.

(Male shown actual size)

Eufriesea auripes

BODY LENGTH
Female 0.71–0.79 in. (18–20 mm)
Male 0.71–0.79 in. (18–20 mm)

FOOD PLANTS Not known.

HABITAT RANGE Locally common in tropical rainforest of the Amazon basin.

Description *Eufriesea auripes* is an orchid bee, a member of the family Apidae in the tribe Euglossini. The male is dark green with iridescent purple highlights, and yellow bristles on the hind legs. In most orchid bees only the male has been identified systematically. The females simply have not been recorded and studied, not least because there are many species and the females look very similar.

Behavior & life cycle Comparatively little is known about the biology of these bees. They are thought to produce only one brood a year, and to be active only during the short rainy season. Some of the bees in this tribe are solitary, while others live in communal aggregations, but they are not truly social bees. In the genus *Eufriesea*, nests are a series of cylindrical cells of resin combined with bark fragments. They may be branched, and they are situated in sheltered dry cavities or protected crevices, usually elevated, such as in rock crevices or tree hollows. *Eufriesea auripes* males collect methyl salicylate (oil of wintergreen) from plants, carrying it in sacks on the hind legs. This is typical of orchid bee males, which gather aromatic chemicals from bark, decaying wood, or fungi. Exactly how these chemicals are used is unclear, but they may be used in mating.

(Female shown actual size)

Domino Cuckoo Bee *Thyreus lugubris*

BODY LENGTH
Female 0.51–0.55 in. (13–14 mm)
Male 0.51–0.55 in. (13–14 mm)

FOOD PLANTS As a cuckoo bee, this species does not collect pollen for its young, but relies on the provisions gathered by its host.

HABITAT RANGE Locally common, anywhere its host, the Teddy Bear Bee (*Amegilla bombiformis*) is found.

Description A black bee with a distinctive spotted pattern of white hairs on the head, thorax, abdomen, and legs. The bees possess neither scopae nor pollen baskets (corbiculae), because they do not forage for pollen.

Behavior & life cycle Active from December to April, this cuckoo bee (family Apidae, tribe Melectini) relies on its host to provide a nest for its young. Typical of cuckoo bees, the mated female locates the host's nest by its scent. A domino cuckoo female spends less than eight minutes in the vicinity of the host's nest. She waits until the host female has left the nest before entering, and her visit into the nest takes a few minutes. She lays two to four eggs in a clump on the pollen and nectar stores in the brood chamber. Consequently the cell contains several cuckoo bee larvae, which hatch sooner than the host bee larvae. They consume the stores, developing rapidly in the egg chamber. The cuckoo larvae vary in size, possibly due to hatching order—those that hatch first eat most stores and are more likely to become fully developed adults. The cuckoo larvae are more active than the host bee larvae, and because they hatch first they are bigger than the host larvae so are better able to eat the food supplies. Consequently, the host larvae either starve or emerge as malnourished and stunted adults.

(Female shown actual size)

Colletes cunicularius

BODY LENGTH
Female 0.55–0.59 in. (14–15 mm)
Male 0.55–0.59 in. (14–15 mm)

FOOD PLANTS They preferentially pollinate willows (*Salix*), but also many other plants. Males are the exclusive pollinators of the orchids *Ophrys exaltata* and *O. arachnitiformis*.

HABITAT RANGE Generally widespread, but rare in the British Isles. Sandy areas, both coastal and inland.

Description *Colletes cunicularius* is similar in shape to a western honey bee. It is a dark brown bee with golden-brown fur on the head, thorax, and abdomen, and pale bands across the abdomen. The male has long pale hairs on his face.

Behavior & life cycle This is a member of the plasterer or cellophane bee group (family Colletidae), so named because they line their nest burrows with a natural polyester, described as a laminester. This coating waterproofs the chamber in which the egg is laid. With an annual life cycle, the adults are only seen flying out in early spring to forage, build their nests, and mate before laying eggs that will eventually develop into adult bees that emerge the following year. After mating, the female digs an underground nest with egg chambers. She laminates the egg chambers, or cells, and provisions them with pollen and nectar stores before laying her eggs. The nests may be made singly or in groups of individual nests aggregated in a "bee village" where conditions are suitable. These bees may therefore be encountered in large groups. In mainland Europe this species is a host of the parasitic bee *Sphecodes albilabris*. In Britain there is a distinct subspecies, *C. c. celticus*, which has Dufour's gland secretions that are significantly different from those produced by the bees of mainland Europe.

(Female shown actual size)

Colletes succinctus

BODY LENGTH
Female 0.31–0.51 in. (8–13 mm)
Male 0.31–0.47 in. (8–12 mm)

FOOD PLANTS Pollen is collected from heaths and heathers (*Erica* and *Calluna*), and from yellow Asteraceae flowers. Nectar is gathered from sweet clover (*Melilotus* spp.), yarrow (*Achillea millefolia*), and creeping thistle (*Cirsium arvense*).

HABITAT RANGE Widespread and common on coastal and inland dry heath and moorland.

Description *Colletes succinctus* is similar in shape to the western honey bee, but slightly smaller, with a dark brown head, thorax, and abdomen, and with distinctive white bands across the abdomen. There is dense red-gold fur on the head and thorax.

Behavior & life cycle This is another member of the Colletidae family. Active only during high summer, from July to early September, these non-aggressive bees are often seen in large numbers close to the ground. Males patrol for females, and may form a ball of several males surrounding a female, rolling on the ground until one male mates with her. Nest burrows are excavated by females either singly, as seen in southern England, or in large aggregations. For example, in North Yorkshire, UK, one aggregation containing 60–80,000 nests was found along a 330-foot (100-m) length of riverbank. Typically, they are sited in warm, south-facing soil or earth banks that are bare or sparsely covered with plants. The nest contains five to six cells, which the female waterproofs by lining them with a cellophane-like polymer. She provisions each cell with a liquid mixture of honey and pollen upon which a single egg is laid. The developing bee overwinters in its nest cell at any stage of its development from egg to pupa. The new adult emerges in the following year. Only one brood is produced each year.

(Male shown actual size)

Sunflower Sweat Bee *Dieunomia triangulifera*

BODY LENGTH
Female 0.35–0.61 in. (9–15.5 mm)
Male 0.31–0.55 in. (8–14 mm)

FOOD PLANTS Sunflower (*Helianthus annuus*).

HABITAT RANGE Locally common on farmland near sunflowers.

Description A black-brown bee with sparse, pale-colored hairs.

Behavior & life cycle A solitary bee of the family Halictidae that produces only one brood a year, these bees become active as their primary pollen source comes into flower. Sunflowers start growing in May, typically flowering in late summer. In Kansas, these bees compete with eight other bee species, such as *Andrena helianthi*, which also gather pollen from sunflowers during the same period. *Dieunomia triangulifera* has an advantage, however, because it transports pollen both on its back and on its hind legs. Males emerge a few days before the females in mid-August. They patrol large areas in their search for receptive females, who have a distinctive odor produced by a gland in the head. The males can differentiate the scent of the newly emerged and receptive females from that of the mated females. A few days later, the females start excavating their nests in large aggregations (forty nests per 10 square feet/1 m²) in moist, compacted soil in bare open fields, between rows of crops, in field margins, and on roadside verges. This activity continues until early September, but each female makes only one nest. Most pollen is gathered in early morning and late afternoon. Females live, on average, for thirteen days and produce two to six offspring.

(Female shown actual size)

Sweat Bee *Augochlorella aurata*

BODY LENGTH
Female 0.20–0.24 in. (5–6 mm)
Male 0.20 in. (5 mm)

FOOD PLANTS *Opuntia, Polygonum hydropiperoides, Viburnum rufidulum, Rubus,* and *Aster.*

HABITAT RANGE A common North American grassland and urban species, extending from the mountainous regions and central plains to the East Coast lowlands.

Description The female is bright green to yellow-green or coppery-green, with golden-white hair on the back of the head, thorax, legs, and abdomen and white hair on the face and front of the body. The male is brilliant green or coppery red.

Behavior & life cycle This primitively eusocial bee (family Halictidae, tribe Augochlorini) is active from May to November. The mated female or foundress builds a nest, often in an aggregation, in a grassy embankment or south-facing slope, and provisions the cells for her first brood. On average, seven eggs are laid. The female young emerge as workers, and they are mostly sterile. The foundress lays more eggs, relying on her daughters for food and to nurture the developing brood. This second brood develops into a reproductive population of males and females. They emerge in late summer, and mate. Only the mated females overwinter. Rarely, a nest has two small co-foundresses, who produce twice the amount of brood. They divide duties, with one foundress laying eggs while the second forages, despite being a mated female. Should the foundress queen die before the second brood is produced, one of her daughters becomes a replacement queen. Nests produced at the most northerly parts of the sweat bee's range tend to be less social, with few or no workers produced by the mated female.

(Female shown actual size)

Yellow-Footed Solitary Bee *Lasioglossum xanthopus*

BODY LENGTH
Female 0.43–0.47 in. (11–12 mm)
Male 0.43 in. (11 mm)

FOOD PLANTS Sea campion (*Silene uniflora*), bramble (*Rubus fruticosus* agg.), blackthorn (*Prunus spinosa*), clover (*Trifolium* spp.), ground-ivy (*Glechoma hederacea*), knapweed (*Centaurea* spp.), field scabious (*Knautia arvensis*), and dandelion (*Taraxacum officinale*).

HABITAT RANGE Common on chalk grassland, coastal areas, and cliffs.

Description A black bee with white facial hair, bright brown antennae, cloudy wings, golden and white bands across the abdomen. The legs and feet are golden yellow.

Behavior & life cycle *Lasioglossum* is a genus in the family Halictidae, tribe Halictini. It is the largest genus of bees, with some 17,000 highly variable species. It is usual for these species to mate before overwintering. They show varying degrees of sociality. In some species, the foundress raises only daughters in her first (spring) brood. They are unable to mate and become workers, tending the nest and foraging. The foundress may be larger than her daughters. She may produce three or four more broods. Often the last brood is solely male, with reproductive females produced shortly beforehand. However, other members of the genus are solitary. In the case of *L. xanthopus*, females are active from early April to at least August, but males are not seen until August to mid-October, when they mate. The mated females then hibernate through the winter. Nest burrows are difficult to find and are covered by vegetation. One large aggregation of nests has been found, but the few other examples found have been solitary, and the species is not considered to be social. The cuckoo bee for this species is *Sphecodes spinulosis*.

(Female shown actual size)

Wallace's Giant Bee *Megachile pluto*

BODY LENGTH
Female 1.54 in. (39 mm)
Male 0.91 in. (23 mm)

FOOD PLANTS Gathers pollen from several different species, mostly in the family Myrtaceae.

HABITAT RANGE Once thought to be extinct because the species had not been seen for 120 years. A very rare bee of tropical lowland rainforest.

Description The female is velvety black with a band of white fur at the top of her abdomen and large jaws used to collect resin. Her wingspan is 2.48 inches (63 mm).

Behavior & life cycle These bees (family Megachilidae, tribe Megachilini) are probably communal. They nest only in active termite (*Microcerotermes* spp.) nests, with up to six female bees found in a nest. The females gather resin, roll it into a 0.39-inch (10-mm) diameter ball, and, holding it in their jaws, fly to the nest. They also collect bundles of wood fibers, which are combined with the resin to produce a mixture that sets black, hard, and waterproof, sufficiently tough to protect the nest from termites. The nest has a horizontal entrance tunnel and a main vertical shaft that is wide enough for two females to pass one another. Brood cells are placed horizontally off the main shaft. In a nest found in 1981, 157 brood cells were uncovered but only 25 were in use. Brood cells are reused. A moist mass of pollen and nectar is deposited in the base of each cell, with an egg, 0.35 inches (9 mm) long, laid on top. The females collect resin to feed their young. Males are territorial near the resin source and the nest. They remain vigilant, resting on a vine and flying rapidly to pursue intruders or females, with whom they will mate.

(Male shown actual size)

Alfalfa Leafcutter Bee *Megachile rotundata*

BODY LENGTH
Female 0.31–0.35 in. (8–9 mm)
Male 0.28–0.31 in. (7–8 mm)

FOOD PLANTS A wide range of plants including blueberry (*Vaccinium*), Asteraceae, Boraginaceae, Lamiaceae, Rosaceae, Verbenaceae, and Fabaceae.

HABITAT RANGE Common and widespread in a wide variety of habitats. Introduced to Australia, USA, Canada, and New Zealand.

Description The female is a black bee with a narrow face and yellowish-white hair and narrow yellow-white hair bands. The scopae are pale, and she is distinctly white beneath the scopae. The male is black-brown with yellowish front feet and yellow hair on his face.

Behavior & life cycle *Megachile rotundata* is a member of the Megachilidae family (tribe Megachilini). They are active in high summer, from June to August, producing one and sometimes two broods. The males emerge one to three days before the females. They can mate with many females, but each females mates only once, soon after emergence. The females must eat pollen for their eggs to mature. Then they start building and provisioning nest cells. They nest in pre-existing cavities, lined and partitioned with pieces of leaf, cut in scissor-fashion using their jaws. The female chews the edge of each leaf until it is sticky, which glues it in place. She places pollen and nectar at the bottom of the nest cell, supplying on average 17 percent less of these stores for her male offspring. She then lays an egg and seals the cell. She lives for seven to eight weeks, and will lay an average of fifty-seven eggs. The bees overwinter as prepupae held in diapause. *M. rotundata* suffers from a number of pests and diseases, most commonly chalkbrood (*Ascosphaera aggregata*).

(Male shown actual size)

Megachile integra

BODY LENGTH
Female 0.43–0.51 in. (11–13 mm)
Male 0.43–0.47 in. (11–12 mm)

FOOD PLANTS *Erigeron, Galactia, Glycina, Koellia, Phaseolus, Sesbania macrocarpa,* and *Strophostyles* spp..

HABITAT RANGE Locally common in a wide variety of habitats, including moist sand prairies, moist meadows in woodland areas, thickets, fens, swamps, and rocky bluffs.

Description The female is black with dense whitish hair on her face and cheeks. The thorax has yellow-white fur with some black hairs. She has a rather dense fringe of very short dark scopal hairs and creamy-white scopa. The male is black with reddish-yellow legs, and his facial hair is yellow with a dense transverse brush of elongate yellowish hairs across the upper face, with short white fur on the cheeks and dense yellow-white fur on the thorax. The male's abdomen has dense yellow hairs, and a wide band of black hairs.

Behavior & life cycle This species belongs to the family Megachilidae (tribe Megachilini). Active from March (in the south) to September, the mated females nest in sandy soil in an aggregation, with nests sometimes 3 feet (1 m) apart. A small mound of excavated soil is left near the entrance. The main shaft descends at 45 degrees, and the burrow is 2.5–5.5 inches (6–14 cm) deep, with a single nest cell at the end. Bramble (*Rubus fruticosus*) leaves, for example, are cut and used to form uniform nest cups. The nest is scented with short-chain fatty acids produced by the Dufour's gland; these chemicals are found in the pollen provision for the brood and may improve their storage. Nests may be predated by fire ants (*Solenopsis invicta*).

(Female shown actual size)

Mason Bee *Osmia cornuta*

BODY LENGTH
Female 0.39–0.59 in. (10–15 mm)
Male 0.39–0.59 in. (10–15 mm)

FOOD PLANTS Visits a wide variety of plants for pollen and nectar, especially fruit trees such as apple (*Malus* spp.), almond (*Prunus amygdalus*), plum (*P. domestica*), and pear (*Pyrus* spp.).

HABITAT RANGE Common in orchards, gardens, and allotments.

Description The female is a black bee with an orange abdomen and feet, black fur on her face and thorax, and long red-brown hairs on the abdomen. The males are similar but have white fur on the face and a black abdomen with long orange hairs.

Behavior & life cycle Like most mason bees (family Megachilidae, tribe Osmiini), this species prefers to nest in pre-existing cavities. Males emerge first. A newly mated female may build up to seven nests with a series of brood cells separated by mud partitions. Female eggs are laid first and provided with larger stores of provisions of nectar and pollen. Eggs hatch a few days after they are laid in February/March but the larvae are not ready to pupate until August/September. They reach the adult stage but do not emerge. Instead they go into diapause and spend the winter in their cells. *Osmia cornuta* and *O. bicornis* nests are often invaded by pests and predators. Examples include the fly *Cacoxenus indagator*, which lays its eggs on the pollen masses. Their larvae consume the bee larvae. Acarine mites (*Chaetodactylus osmiae*) compete with the bee larvae for food. The malnourished emerging adult bees are stunted. The fly *Anthrax anthrax* kills the pupae, destroying the nest. The beetle *Trichodes apiaris* lays its eggs in the nests; their larvae eat any bee eggs and larvae.

(Male shown actual size)

Red Mason Bee *Osmia bicornis*

BODY LENGTH
Female 0.39–0.59 in. (10–15 mm)
Male 0.39–0.59 in. (10–15 mm)

FOOD PLANTS Visits a wide variety of plants, especially almond (*Prunus amygdalus*), raspberry (*Rubus idaeus*), plum (*Prunus domestica*), pear (*Pyrus* spp.), and apple (*Malus* spp.).

HABITAT RANGE Common and widespread in rural and urban environments, including parks and gardens.

Description The bee is greenish-black to bronze. The head, thorax, and abdomen have pale brown hairs. The top of the thorax has red or red-brown fur, while fur on the abdomen, pollen scopae, and legs is orange-brown. The female has horns on her face.

Behavior & life cycle *Osmia bicornis* is a member of the Megachilidae family (tribe Osmiini). They are active from March to June, nesting in soft mortar, soil, dead wood, gaps under roof tiles, or existing burrows. They also nest readily in straws, stiff cardboard tubes, and cut bamboo stems placed 3–6.5 feet (1–2 m) above the ground. One brood is raised each year. The males emerge about two weeks before the females. The mated female prepares her nest by lining it with pellets of soft mud carried in her jaws. She uses her horns to smooth it onto the tunnel walls. She provisions the cell with pollen and honey, kneading it into a paste using her jaws. The slightly curved egg, 0.2 inches (4.5 mm) long, is laid on top of this paste, and each completed cell is sealed with mud. She normally lays four to six eggs. The male eggs are laid last, towards the entrance of the tunnel. She plugs the completed burrow with a spiral of mud. The young bees hatch and develop in their nest cells for the rest of the summer, overwintering as fully formed bees in brown silken cocoons.

(Male shown actual size)

Violet Carpenter Bee *Xylocopa violacea*

BODY LENGTH
Female 0.79–0.94 in. (20–24 mm)
Male 0.79–0.83 in. (20–21 mm)

FOOD PLANTS Visits a wide range of pollen and nectar sources, including *Centaurea, Lathyrus, Narcissus, Prunus, Lamium, Lonicera, Vicia faba,* and *Wisteria sinesis.*

HABITAT RANGE Common and widespread in forest, woodland, grassland, nature reserves, and gardens.

Description These large bees, belonging to the family Xylocopinae, are black with dark wings, which in bright light shine purple or blue. They look brown in poor light. The abdomen is hairless. Females have black antennae but on males antennae have two orange-red segments near the tip. Males are smaller, and occasionally dwarf.

Behavior & life cycle A nectar robber, *Xylocopa violacea* is active from February to June, raising one and sometimes two broods each year. They nest from April in dead wood, grass stems, or bamboo canes. They also nest in wooden houses and are sometimes regarded as pests. The males emerge first. After a few days of foraging, they patrol several times a day in search of females. Some males are territorial and will pursue any intruders for a distance of 135–165 feet (40–50 m). Should a male encounter a female during his patrol, he will mate. Males and females may both have multiple matings. The mated female excavates a tunnel filled with a line of brood cells. She completes one cell before starting the next. She makes a pollen and nectar paste, onto which she lays a single egg, 0.4–0.5 inches (9–12 mm) in length. She seals the chamber with saliva. Females defend their nests, emitting intermittent buzzes or flying at the intruder. They are not aggressive to humans and rarely sting.

Bumble Bees (Genus *Bombus*)

Bumble bees are all members of the family Apidae and are placed in the tribe Bombini. They are social bees found throughout the northern hemisphere and South America. There are about 250 species worldwide. Most species are temperate, although a few are tropical. Adapted to a temperate climate, bumble bees are large and furry, and are capable of flying in cool conditions.

Bumble bees live in colonies for part of the year. Queens overwinter, generally in a hole in the ground. They emerge in spring and search for a suitable nest site, which varies with the species. The queen then forages for nectar and pollen, and settles down in the nest. She secretes wax and fashions it into cells where she lays eggs. She feeds the newly hatched larvae with nectar and pollen. These larvae pupate and emerge as sterile workers. The first workers to emerge are often very small. As the workers take over foraging and nest duties, the queen concentrates on egg laying and the colony expands rapidly. The workers raised later are larger.

The maximum colony size depends on the species, varying from tens of workers to several hundred. In late summer, a "switch point" occurs, and eggs laid thereafter become new queens and males (drones). The queen's fertile offspring leave the nest and mate with others. In general, queens only mate with a single male, but multiple mating occurs in some species. Following the emergence of these new breeding adults, the colony declines, leaving only the newly mated queens to overwinter and repeat the cycle. The old queens, workers, and drones die. Generally only one nest cycle occurs per year, but in some species, in some

Above *Bumble bees are a common sight in our gardens and are beneficial to both gardeners and growers because they buzz pollinate flowers.*

situations, two cycles may occur, and in exceptional circumstances perennial colonies can occur.

Bumble bees fall into two groups in terms of their method of pollen feeding. Some, such as *B. terrestris*, are termed "pollen storers," as they store pollen in separate cells in the nest and the developing young are fed by the queen or workers. Others, such as *B. pascuorum* and *B. hortorum*, are termed "pocket makers," because they place pollen in pockets built into the side of the brood cells. Their developing larvae feed themselves directly. This affects the ease with which bumble bees may be reared in captivity: the pocket makers are more difficult to rear, because it is problematic to place the pollen on the brood cells.

Naturalists have managed to keep bumble bee colonies in artificial homes for several centuries. In recent decades, however, techniques have been developed that permit the commercial production of several species for pollination, especially of greenhouse crops. This has rapidly become a huge global trade. Bumble bees are especially useful for pollination, because they have the ability to "buzz pollinate" flowers such as tomatoes. This involves the bees landing on flowers and rapidly vibrating their flight muscles, sonicating the flower with an audible buzz causing pollen to be released (see page 49).

Bumble bees vary in aggressiveness. Bees from some of the species that form larger colonies can defend their nests tenaciously. The bumble bee's sting is unbarbed, so can be withdrawn easily from human skin, but a sting may nonetheless result in a reaction similar to that caused by honey bees or wasps.

(Worker shown actual size)

Garden Bumble Bee *Bombus hortorum*

BODY LENGTH
Queen 0.67–0.79 in. (17–20 mm)
Worker 0.43–0.63 in. (11–16 mm)
Male 0.55–0.59 in. (14–15 mm)

FOOD PLANTS Long-tongued, specializing in red clover (*Trifolium pratense*), foxglove (*Digitalis purpurea*), honeysuckle (*Lonicera periclymenum*), and many other plants.

HABITAT & STATUS Fairly common in a wide range of habitats (gardens, farmland, grassland, and woodland), but seldom found in marshy or upland moorland sites.

Description A black bumble bee with three lemon-yellow stripes (two on the thorax, one on the abdomen) and a white tail. Male, queen, and worker all share this coloration, although the male has a yellow head. Only rarely are all-black bees found. This bee is distinguished from similar species by its long, narrow face. Its very long tongue (up to 0.79 inches/20 mm) often protrudes as the bee flies between deep flowers such as foxgloves. Older workers often have a shiny thorax, where the hair has rubbed off against the nest.

Behavior & life cycle *B. hortorum* form short-lived colonies of up to a hundred workers during the summer, nesting in old small mammal nests, shallow scrapes, under tree roots etc., but no deeper than 20 inches (500 mm) below the surface. These bees will also nest in man-made locations such as in buckets, under sheds, or in compost bins. The queen provisions her nest with supplies, including honey pots of stored nectar and pollen. Males patrol an area, no more than 3 feet (1 m) above ground, scenting it with pheromone to attract the young queens as possible mates. This bee's nest is parasitized by the cuckoo bee *Bombus barbutellus*. *B. hortorum* is a placid bee, more likely to wave its mid-leg as a warning than to sting. The species has been introduced to New Zealand and Iceland to assist with crop pollination.

(Queen shown actual size)

Tree Bumble Bee *Bombus hypnorum*

BODY LENGTH
Queen 0.71 in. (18 mm)
Worker 0.55 in. (14 mm)
Male 0.63 in. (16 mm)

FOOD PLANTS A short-tongued species that visits *Rhododendron*, grape hyacinth (*Muscari* spp.), fruit trees, and soft fruit.

HABITAT & STATUS A lowland bee, common throughout central Europe and Asia. Its range has grown to include England in 2001 and Iceland in 2008. Open woodland, grassland, urban parks and gardens, and roadside verges.

Description Male, queen, and worker all have a ginger-brown furry body with a black head and a white tip to the tail. The male may have some yellow hair on his face.

Behavior & life cycle A social bee with a short spring colony life, nesting in cavities above ground, in colonies of 150 or more workers. It will often use bird boxes, but also roof spaces and tree holes. The queens emerge in spring (February to early March), with the worker eggs being laid, developing into adults, and flying out to forage shortly afterwards. Tree bumble bees are pollen storers, with the queen and workers gathering pollen and storing it in pollen cells from which it is fed to the developing larvae. The species is most visible in May and June, when the males emerge to mate with young queens—although in some areas a second burst of brood activity is possible. Male bees are seen hanging about the entrance to the nest, apparently waiting to mate with the virgin queens as they leave the colony. This species has developed a reputation for being aggressive, but this may because it often nests in situations close to domestic activity, so that there is a strong likelihood of coming into contact with humans, leading to stinging incidents. Like most bumble bees, it is likely to sting only if provoked.

(Queen shown actual size)

Common Eastern Bumble Bee *Bombus impatiens*

BODY LENGTH
Queen 0.83 in. (21 mm)
Worker 0.63 in. (16 mm)
Male 0.67 in. (17 mm)

FOOD PLANTS Visits a wide range of wild flowers and garden plants; typically seen on goldenrod (*Solidago*) in late summer and fall.

HABITAT & STATUS A common generalist bee of many rural and urban habitats, from the cold of Canada and Minnesota to subtropical Florida. Spreading westwards and southwards in the USA.

Description A black-brown bee with golden fur on the thorax and black hairs on the head, abdomen, and legs. Worker bees closely resemble the queen but are smaller. The male has some golden hairs on his face but is otherwise like the other castes.

Behavior & life cycle This eusocial bee has very large colonies and nests above and below ground, often using disused rodent holes or grassy tussocks. It has a long flying season, typically March to November, but is seen as early as January and February in Florida, where it has a short or minimal overwintering period. They pollinate outdoor crops such as soybean, sunflower, and field beans, as well as fruit and nut crops. *B. impatiens* has been successfully reared in captivity and is now used commercially for greenhouse pollination of squashes, peppers, tomatoes, etc. in California and Mexico, far outside its native range. In the west it has replaced the previously used western species *B. occidentalis*, because most wild and commercial populations of *B. occidentalis* disappeared after this species was developed for use by the bumble bee industry. Efforts are under way to obtain permits authorizing the use of exotic *B. impatiens* for outdoor field pollination in California, where a very similar and very closely related species, the California native *B. vosnesenskii*, is abundant.

(Queen shown actual size)

Common Carder Bee *Bombus pascuorum*

BODY LENGTH
Queen 0.63–0.71 in. (16–18 mm)
Worker 0.39–0.59 in. (10–15 mm)
Male 0.51–0.55 in. (13–14 mm)

FOOD PLANTS Legumes, vetches, clovers (*Trifolium* spp.), brambles (*Rubus* spp.), mullein (*Verbascum* spp.), deadnettles (*Lamium* spp.), thyme (*Thymus* spp.), sage (*Salvia* spp.), and lavender (*Lavendula*).

HABITAT & STATUS Fairly common in a wide range of lowland and upland habitats, but rarely seen in wet and marshy areas.

Description This species is almost entirely brown or ginger. Some individuals have lighter abdomens, but there are always a few black hairs on the abdomen. There are regional color differences: in some places, the bees are brown whereas in another locality they may have red-brown fur. All three castes look similar, but the queen is slightly larger and the males lack pollen baskets on their hind legs. Whereas other bumble bees often go bald on the top of the thorax, this is rarely seen in *B. pascuorum*—but the ginger-brown fur often becomes pale as the bees age.

Behavior & life cycle The common carder bee forms medium-sized colonies with approximately 60–150 workers. The nest is longer-lived than that of many bumble bees, lasting up to twenty-five weeks. At the northern limit of this bee's range the stages of colony life are up to two months behind colonies in the southern part of the range. *B. pascuorum* may nest above ground, sometimes in holes in trees or bird boxes, but most commonly uses old nests of small mammals such as mice or voles, or nests in tussocks of long grass, under hedges, or in leaf litter. Like other carder bees, moss and grasses are gathered and used to line and cover the nest. The cuckoo bee *Bombus campestris* parasitizes the nest.

(Queen shown actual size)

Broken-Belted Bumble Bee *Bombus soroeensis*

BODY LENGTH
Queen 0.63 in. (16 mm)
Worker 0.47 in. (12 mm)
Male 0.51 in. (13 mm)

FOOD PLANTS Clovers (*Trifolium* spp.), ling (*Calluna vulgaris*), bellflowers (*Campanula* spp.), scabious (*Scabiosa* spp.), and bird's-foot trefoil (*Lotus corniculatus*).

HABITAT & STATUS A bee of moorland and heathland habitats, but also other environments including chalk grasslands. Scarce in the UK, absent from Ireland.

Description A thickly furred and vividly striped yellow and black bee with a white tail, recognized by the central narrowing of the broad yellow stripe on the abdomen—but it is notoriously difficult to distinguish this species from similar-looking bumble bees. Coloration is very similar across queen, worker, and male bees, although the queen is noticeably larger than the others in the colony, and males have a narrow band of deep orange between the black of the lower abdomen and the white of the tail.

Behavior & life cycle *B. soroeensis* lives in small colonies of up to a hundred individuals. A queen is the only member of the colony to survive the winter. In late summer she mates with a male before hibernating in a burrow underground though the winter. In spring she makes a subterranean nest, usually in an old mouse or vole nest, and lay eggs which hatch into the colony's first worker bees. When cold weather arrives in late fall, the queen will produce new queens and males before the other bees in the colony die off; the new queen bees mate, and the full cycle repeats. As in other bumble bees, the males have no sting; workers can sting, but can withdraw their stings after use and thus survive.

(Queen shown actual size)

Short-Haired Bumble Bee *Bombus subterraneus*

BODY LENGTH
Queen 0.79–0.87 in. (20–22 mm)
Worker 0.47–0.71 in. (12–18 mm)
Male 0.59–0.63 in. (15–16 mm)

FOOD PLANTS Long-tubed flowers including red clover (*Trifolium pratense*), honeysuckle (*Lonicera periclymenum*), *Rubus*, thistles, mustard, and *Echium*.

HABITAT & STATUS Flower-rich grassland and heathland, and possibly low-intensity farmland, scrubland, and marshland. Widespread but rare.

Description With distinctive even short hair, this bee has two color forms. The darker form *(B. s. subterraneus)* is confined to southern Scandinavia, Switzerland, northern Italy, and southeast France. The light-colored subspecies (*B. s. latreillellus*) can be found wherever the darker form is not present.

Behavior & life cycle *Bombus subterraneus* nests underground, usually in old mouse nests. It is thought to form large colonies, but these have rarely been observed. Hibernation sites for queens are unknown. This was one of four species of bumble bee introduced into New Zealand from the UK in the late nineteenth century for pollination of red clover. Although now rare in New Zealand, it still survives there. In contrast, the last recorded sighting in the UK was in Dungeness, Kent, in 1988, and it was subsequently declared extinct. In 2009, a pioneering project to reintroduce it to the UK was initiated, and the New Zealand bees—as well as others from Sweden—have been investigated as source material. In 2013, several batches of Swedish queen bees were reintroduced to the UK. Newly hatched queens were observed looking for mates so it is hoped that they have mated and will overwinter and multiply.

(Queen shown actual size)

Buff-Tailed Bee *Bombus terrestris*

BODY LENGTH
Queen 0.79–0.87 in. (20–22 mm)
Worker 0.43–0.67 in. (11–17 mm)
Male 0.55–0.63 in. (14–16 mm)

FOOD PLANTS A short-tongued species that visits a wide range of flowers but prefers open single flowers and composites.

HABITAT & STATUS Common and wide-ranging, especially in lowland areas, from gardens and parkland to farmland, heath, and woodland glades. Less common in upland or marshy habitats.

Description This bumble bee has two dark yellow bands, one on the thorax near the head and the second on the abdomen. Workers have a white tip to the tail, and the queen has a buff-white (or occasionally orange) tail tip.

Behavior & life cycle The queens emerge from February onwards. The nest becomes large, with up to five hundred bees, and is typically sited underground in an old rodent hole, with an entrance tunnel up to 6 feet (2 m) long. The workers fly for much of the year, with males emerging to mate with the young queens from July to October. Usually only the newly mated queen overwinters, but some colonies have remained active throughout recent warmer winters in southern England (and similar latitudes), with worker bees seen flying during January. It was the first bumble bee species to be reared commercially, and it is now used worldwide to pollinate greenhouse crops such as tomato, eggplant (aubergine), zucchini (courgette), watermelon, bell pepper, and strawberry. Outdoors, it is an important pollinator of fruit and crops such as alfalfa, onion, rape, and sunflower. Commercial use has led to this bee escaping and colonizing Tasmania and South America. It was deliberately introduced to New Zealand in the late 1800s. *B. terrestris* is parasitized by the cuckoo bee *Bombus vestalis*.

Stingless Bees (Tribe Meliponini)

Stingless bees are distributed through the tropics in Australia, Asia, Africa, and the Americas (where 75 percent of species are found). Like honey bees and bumble bees, they are members of the Apidae family, but they belong in a separate subdivision, the Meliponini tribe. There are over five hundred known species, and many more remain to be described. Considering their diversity and economic importance, the stingless bees are understudied.

It is believed that these bees evolved on the Gondwanan continent some one hundred million years ago, long before honey bees first appeared. The earliest known example of a stingless bee is *Cretotrigona prisca*, which has been found fossilized in amber dating back sixty-five million years. Having evolved apart from other bees, the stingless bees display several distinctive adaptations and features. Like certain solitary bees in Australia, they have evolved independently to fit niches that elsewhere in the world are occupied by honey bees and bumble bees.

Meliponini females have residual and futile stingers, hence the term "stingless." However, stingless bees are not defenseless, and will bite potential intruders. They are truly social bees, forming permanent colonies with a single queen, sterile female workers, and males. The queens mate only once. They do not migrate or swarm to reproduce but must first build a new nest.

Like honey bees (Apini), the workers have pollen baskets or corbiculae rather than scopae. Stingless bees are usually generalists, visiting a wide array of plants as they gather provisions for the hive. When seeking forage, these bees communicate where to find the best supplies, but the waggle dances seen in *Apis mellifera* have not been observed in stingless bees. They produce less honey than the honey bees and store their honey in pots within the nest. Indeed, stingless bee honey is often called "pot honey." Increasingly, these bees are being exploited commercially to pollinate crops such as coffee, coconut, mango, and macadamia nuts, among many other plants.

Below *Stingless bee nests often have guards who defend the entrances by biting and buzzing at any potential invaders.*

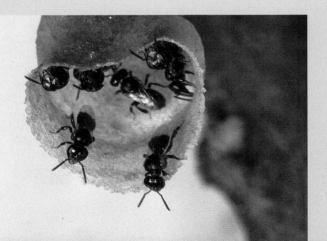

(Male shown actual size)

Xunan Kab Bee or Royal Lady Bee *Melipona beecheii*

BODY LENGTH
Queen length unknown
Worker 0.39 in. (10 mm)
Male 0.35 in. (9 mm)

FOOD PLANTS Visits a wide range of plants including Apocynaceae, *Cassia emerginata, Cordia gerascanthus, Dalbergia reteusa, Melanthera* spp., and *Solanum* spp.

HABITAT & STATUS Tropical and subtropical lowland rainforest and dry forest. Threatened (very rare).

Description A brown bee with golden bands across the abdomen and reddish brown legs. They have pale gray fur on the head, thorax, and legs. Hair is sparse on the abdomen.

Behavior & life cycle These bees nest in tree cavities in mature forest. They are also kept in hollow logs or man-made boxes. Within the cavity, they make stacks of dark wax comb. Storage pots for honey and pollen are built on the edges of the nest, attached to the wall of the hive. They are sealed when full. They do not feed their larvae, but provision the brood cells with pollen and honey stores. The queen lays an egg on top of the stores and the cell is sealed. Worker bees help the young bee to emerge, cutting away the wax capping so that the young bee can climb out from its papery cocoon. Workers are specialists when foraging. Even though nectar and pollen foragers make the same number of flights, their life spans differ. Nectar gatherers were observed to be active all day but to die after three days, whereas pollen gatherers only worked for one to three hours per day and lived for twelve days. This bee buzz-pollinates flowers. When the food pots or larval cells in the nest have been disturbed, the phorid fly *Pseudohypocera kerteszi* will invade colonies. Colonies have 500–2,500 bees. They are not aggressive.

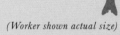
(Worker shown actual size)

Nannotrigona testaceicornis

BODY LENGTH
Queen 0.18 in. (4.5 mm)
Worker 0.16 in. (4 mm)
Male 0.16 in. (4 mm)

FOOD PLANTS A wide variety of plants, including the Lamiaceae (e.g. Salvia) and commercial crops such as strawberry (*Fragaria*).

HABITAT & STATUS Quite common and widespread in a wide range of tropical habitats, including cities.

Description A small black bee with gray hair.

Behavior & life cycle Forming colonies of 2,000–3,000 bees, *N. testaceicornis* nests in groups of colonies in holes in trees or in man-made structures. The short nest entrance is made of wax mixed with resin. Guarded by day, the entrance is closed at night with a curtain of loosely knit resin and wax–resin mixture. These bees are unaggressive, and the workers hide when disturbed. Sometimes this species shares a mixed hive with a larger and very aggressive stingless bee, *Scaptotrigona depilis*. If disturbed, *S. depilis* will attack a human as far as 6 feet (2 m) from the hive, becoming tangled in the hair and biting with its mandibles. The invading *N. testaceicornis* worker bees modify the entrance and gather food and building materials for the host colony. The hosts tolerate the invading bees, although sometimes they bite the invaders' wings. Nest sharing is seen in other stingless bees, usually between closely related species. In the case of *N. testaceicornis* and *S. depilis*, both species gather chemicals from the orchid *Mormolyca rigens* for communication, so the chemical messages of the two species did not conflict. *N. testaceicornus* is used commercially to pollinate cucumbers, sweet peppers, and strawberries in greenhouses.

(Worker shown actual size)

Sugarbag Bee *Tetragonula carbonaria*

BODY LENGTH
Queen 0.28 in. (7 mm)
Worker 0.16 in. (4 mm)
Male 0.16 in. (4 mm)

FOOD PLANTS Seen on a wide range of native plants including *Eucalyptus, Callistemon, Melaleuca,* cycads, *Banksia, Protea,* and orchids, plus fruit trees and non-native garden plants such as sunflowers.

HABITAT & STATUS Common in gardens, community gardens, orchards, forest, and coastal regions.

Description The jet-black females have sparse hair on the top of the thorax and short dense hair on the sides of the thorax. The males look similar, but are dark red-brown.

Behavior & life cycle The nest is made of wax and resin, with the brood cells set in a spiral comb. It is built in a tree cavity or a man-made box. The bees forage within 80 feet (25 m) of the hive, visiting a wide range of plants. These bees deal with the small hive beetle (*Aethina tumida*) and other parasites by coating the invader in batumen, a resin and wax mixture. They only cease their attack when the beetle stops moving; it then shrivels and dies—in effect, it is mummified. They can also defend the nest aggressively. Sugarbag workers may form "fighting swarms," in which thousands die in battle when one colony attempts to invade or rob the nest of another. The battle rarely takes place in the nest itself—thereby protecting the brood and stores—but the defenders form a fighting swarm and take on the invaders some 6 feet (2 m) from the nest. Humans keep these bees in natural logs or hives in towns and gardens. Their honey is marketed as "sugarbag honey." The bees are also used commercially to pollinate macadamias, lychees, melon, mangoes, avocado, and watermelons, and for crops grown inside greenhouses.

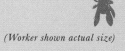

(Worker shown actual size)

Arapuá, Irapuá, or Abelha-Cachorro Bee *Trigona spinipes*

BODY LENGTH
Queen length unknown
Worker 0.24 in. (6 mm)
Male length unknown

FOOD PLANTS A wide variety of plants including the Myrtaceae and the Polygonaceae.

HABITAT & STATUS Locally common in diverse habitats, including savannah and forests in coastal tropical and subtropical regions in South America.

Description A dark black-brown bee, covered in sparse hair, with reddish-brown hind legs.

Behavior & life cycle Colony size ranges from 5,000 to 100,000 workers. The bees build nests on the branches of large trees. Workers cut and add plant material (leaves, bark, flowers) to batumen, a mixture of resin, wax, and mud, which is used to make a protective scutellum or hard shell surrounding layers of brood comb and storage pots for honey and pollen. This species is highly aggressive. It bites, buzzes, and flies at intruders, including humans. It can both defend and claim food sources from, for example, carpenter bees, other stingless bees, and hummingbirds. *T. spinipes* uses pheromones to mark trails for other foragers to follow to a food source. By extending its tongue, the bee deposits saliva containing an aromatic ester, laying a trail that is effective for up to twenty minutes. The scent trail rapidly recruits other foragers to gather supplies. Different species of bees use different chemicals for scent marking. This bee also detects the odors deposited by other bees such as *Melipona rufiventris*. They attack these bees and overwhelm them, and then claim the food source. It also takes nectar from some plants, such as passion fruit, and it is considered a pest for damaging crops in its search for suitable plant material to add to its nest structure.

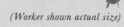
(Worker shown actual size)

Iratim Bee *Lestrimelitta limao*

BODY LENGTH
Queen length unknown
Worker 0.24 in. (6 mm)
Male length unknown

FOOD PLANTS Does not visit plants for food.

HABITAT & STATUS Widespread in the semi-arid desert known as white forest, with cacti, small thorny trees, and scrub.

Description A slender dark black-brown bee with red-brown feet, sparsely hairy on the thorax and legs.

Behavior & life cycle This bee has lost the ability to seek its own food. It has no pollen basket (corbicula) and must parasitize jetaí (*Tetragonisca angustula*) and other bee species, including honey bees. It raids the other bees' nests for pollen and honey, and may also collect plant resins and cerumen (a mixture of wax and resin) from the raided nest. The attack weakens and often kills the invaded colony, due to loss of food stores and losses of larvae and adults. During the raid, *L. limao* releases citral, an alarm pheromone. When raiding bees are killed, yet more citral is released, attracting increasing numbers of robber bees. As the odor pervades the target colony, it also disorientates the host species, leaving them unable to form a cohesive defense. *L. limao* builds its own nest in a natural crevice or cavity, where it stores the raided mixed pollen and honey in pots. It has hollow tubes around the nest entrance, and these blind tunnels seem to be a defense against attack by rival colonies. These bees can fight over potential nesting sites, with battles often costing many robber bees' lives. These battles regulate the density of the local robber bee population.

(Worker shown actual size)

Jetaí *Tetragonisca angustula*

BODY LENGTH
Queen length unknown
Worker 0.12–0.16 in. (3–4 mm)
Male length unknown

FOOD PLANTS Visits a wide range of plants, and is an important pollinator of different plant families.

HABITAT & STATUS Common and widespread in the semi-arid desert known as white forest, with cacti, small thorny trees, and scrub.

Description A slender golden-yellow bee with a black-brown head and thorax. The queen is much bigger, with a significantly larger abdomen.

Behavior & life cycle A colony contains 5,000–10,000 bees. Brood combs are laid in horizontal layers of uniformly sized hexagonal comb. The egg cells are mass-provisioned before the egg is laid and the cell sealed. There are at least four castes: one male and three females—queen, worker, and soldier (or guard). A waste-removing caste has also been discovered, smaller than the guard bees but larger than the workers. Guard bees comprise 1–2 percent of the colony. Some hover near the nest while others stand on the wax nest entrance. There are twenty to forty guard bees at or near the entrance at any one time. The guards have longer hind legs and are 30 percent heavier than a foraging worker bee. They are better able than the workers to withstand challenges by the larger robber bee, *Lestrimelitta limao*, that invades their nests. The guards aggressively defend the nest against any intruder, grabbing and clamping attackers with their jaws. Invasions by *L. limao* can result in colony death, and trapping a scout bee may prevent a potential invasion or give the colony time to mount their defenses.

Honey Bees (Genus *Apis*) 🐝

Honey bees of the genus *Apis* are truly social bees, and arguably occupy the top of the bee evolutionary tree. They form large perennial colonies consisting of a single queen, who may live for several years, male drones present for the summer season, and many sterile workers whose life span varies from a few weeks to many months. These large and complex colonies need large stores of food to survive through periods of shortage or cold weather. All species build combs made of wax secreted by the workers, taking the form of hexagonal cells arranged in one or more vertical combs.

Apis bees are thought to have evolved in Africa before spreading to Europe and Asia, and all but one species are still confined to Asia. The exception, *A. mellifera*, was naturally the most widely distributed, and has now been introduced to all continents except Antarctica. Although most species are tropical, the genus is extremely adaptable, and honey bees live in a wide range of climates from the heat of equatorial Africa to the cold of Scandinavia, and from sea level to the heights of the Himalayas. Honey bees are all members of the family Apidae and are placed in the tribe Apini.

The genus *Apis* comprises three subgenera:
- *Micrapis:* the dwarf honey bees, including *A. florea*
- *Apis:* the medium-sized cavity-nesting species including *A. cerana*, *A. koschevnikovi*, and *A. mellifera*
- *Megapis:* the single species of giant honey bee *A. dorsata*

Two *Apis* species—*A. mellifera* and *A. cerana*—have been kept by humans for thousands of years for their honey, wax, and other hive products. Other species have long been exploited by "honey hunters."

Honey bees are true generalists, capable of exploiting a wide range of different food plants. They have complex pollen-carrying mechanisms with adaptations including pollen baskets, brushes, and combs. Of all bees, *Apis* species possess the most sophisticated and effective communication system for recruiting foragers to food sources, in the form of the representational dance first described by Karl von Frisch (see pages 60–63).

Above *Honey bee queen larvae are fed with royal jelly produced by the workers. This queen cell has been opened to show a larva floating on royal jelly.*

(*Worker shown actual size*)

Asian Honey Bee *Apis cerana*

BODY LENGTH
Queen 0.85–0.89 in. (21.5–22.5 mm)
Worker 0.39–0.75 in. (10–19 mm)
Male 0.39–0.47 in. (10–12 mm)

FOOD PLANTS A generalist species, feeding on plants from many families.

HABITAT & STATUS Widespread throughout south and southeast Asia, occupying a wide range of habitats.

Description Similar to *A. mellifera*, but slightly smaller. Variable in color. There are generally thought to be eight subspecies, two of which are used for beekeeping in India. *A. c. cerana* has black stripes and is mainly found at high altitudes. *A. c. indica* has yellow stripes and is found at lower altitudes, and was originally thought to be a separate species.

Behavior & life cycle Like *A. mellifera*, this species has an advanced dance communication system, but there are slight variations in the dances. Natural nest sites are cavities such as hollow trees. These bees can be kept in wooden hives for honey production, but their natural migratory behavior makes them prone to absconding. *A. cerana* is the natural host to the parasitic mite *Varroa jacobsoni*, as well as other mites, including various *Tropilaelaps* species, and the gut parasite *Nosema ceranae*, which has more recently been found in *A. mellifera*. *A. cerana*, in particular the Japanese bee (*A. c. japonica*), has evolved complex behavior to deal with attacks from various species of hornets such as *Vespa mandarinia*. When a hornet tries to enter the hive, large numbers of bees cluster around the intruder, forming a ball that rapidly heats to 117°F (47°C), killing the hornet but without harming the bees. Other behaviors include "shimmering" at the hive entrance to deter hornets from landing.

(Worker shown actual size)

Giant Honey Bee *Apis dorsata*

BODY LENGTH
Queen 1.20–1.26 in. (30.5–32 mm)
Worker 0.87–0.98 in. (17–25 mm)
Male 0.67–0.79 in. (17–20 mm)

FOOD PLANTS A generalist bee, feeding on plants from many families.

HABITAT & STATUS A common species in southeast Asia, found in a wide range of habitats.

Description Similar to other honey bee species but much larger, and with a relatively long abdomen. Color variable, but many are brownish with black stripes. There are thought to be four subspecies, somewhat variable in size. The largest, *A. d. laboriosa*, originally thought to be a separate species, lives in Bhutan, India, Nepal, and the Chinese province of Yunnan, and at altitudes of between 8,250 and 10,000 feet (2,500–3,000 m) in the Himalayas.

Behavior & life cycle *A. dorsata* forms large colonies with a single comb. The comb is completely covered by bees, which perform defensive "waving" when challenged. Nests are built under overhangs of cliffs, in branches of trees, or under eaves of buildings. This is an aggressive species that cannot be managed in hives, but it has long been exploited by honey hunters. A variant in the forests of Vietnam is "rafter beekeeping," whereby colonies are kept on "rafters" made of wood fixed to poles. A. dorsata has sophisticated seasonal migratory behavior to avoid periods of dearth. It is a major pollinator in the rainforests of Peninsular Malaysia, but unfortunately it is locally threatened by the harvesting of its combs by honey hunters. Although this is a traditional activity and often regarded as "sustainable," whether this is in fact the case clearly depends on the frequency of the harvesting.

(Male shown actual size)

Dwarf Honey Bee *Apis florea*

BODY LENGTH
Queen 0.67–0.71 in. (17–18 mm)
Worker 0.57–0.61 in. (14.5–15.5 mm)
Male 0.31–0.47 in. (8–12 mm)

FOOD PLANTS A generalist species, feeding on plants from many families.

HABITAT & STATUS A common and widespread species from northeast Africa to Malaysia, found in a wide range of habitats.

Description A reddish bee with black stripes. Relatively small body size.

Behavior & life cycle *A. florea* is one of two species of small honey bees that make up the subgenus *Micrapis*. These species are thought to be the most primitive of the honey bees, living in small colonies, and many aspects of their behavior are the same. A single comb is built in an exposed place such as on the branch of a small tree or in shrubs. A form of dance communication is used to recruit other workers to food sources, but the dance differs from that observed in *A. mellifera*, in that instead of dancing on the vertical face of the comb, the bees perform on the horizontal upper surface of the comb, where it is attached to a branch. The dance is relatively simple, consisting of a straight run pointing directly to the food source. This is thought to be a primitive variant of the complex representational dance observed in other species. *A. florea* is parasitized by mites of the genus *Euvarroa*, distantly related to the *Varroa* mites that parasitize the larger *Apis* species.

(Male shown actual size)

Koschevnikov's Bee *Apis koschevnikovi*

BODY LENGTH
Queen 0.85–0.89 in. (21.5–22.5 mm)
Worker 0.71–0.75 in. (18–19 mm)
Male 0.39–0.47 in. (10–12 mm)

FOOD PLANTS A generalist species, feeding on plants from many families.

HABITAT & STATUS A scarce bee, found only in dense tropical forests of Borneo, parts of Malaysia, Indonesia, and Brunei.

Description Reddish in color, but otherwise broadly resembles *A. cerana*.

Behavior & life cycle *A. koschevnikovi* is restricted to Borneo, and is believed to be an island-adapted variant of *A. cerana*. It is thus notable as a good example of Darwinian island evolution. *A. koschevnikovi* is parasitized by a unique species of parasitic mite named *Varroa rinderi*. This is similar to *V. jacobsoni*, which is found on *A. cerana*, but is genetically and morphologically distinct, and appears to be able to live only on *A. koschevnikovi*, even in apiaries where both species of honey bee are present.

(Queen shown actual size)

Western Honey Bee *Apis mellifera*

BODY LENGTH
Queen 0.96–1.00 in. (24.5–25.5 mm)
Worker 0.81–0.85 in. (20.5–21.5 mm)
Male 0.47–0.51 in. (12–13 mm)

FOOD PLANTS Possibly the most generalist *Apis*, feeding on plants from many families.

HABITAT & STATUS Common and widespread in almost every terrestrial environment, absent only from Antarctica.

Description The color is variable, from black to yellow, but often with orange and black stripes. Size varies only a little between subspecies.

Behavior & life cycle These are social bees, with colonies that may number more than fifty thousand workers at the height of summer. There are twenty-five to thirty distinct subspecies, divided into several groups—generally agreed to be the African lineage (A), the west and north European lineage (M), the southeast Europe lineage (C), the Near and Middle Eastern lineage (O), and a fifth found only in Ethiopia (Y). *A. mellifera* is native to Europe, Africa, and much of Asia, and has been introduced to all continents except Antarctica. It is the most widely used commercial pollinator. The practice of transporting hives from place to place is highly developed in North America, where vast numbers of hives are often moved considerable distances as different crops require pollinating. In the wild, *A. mellifera* nests in cavities such as hollow trees, but it is very adaptable and will occupy a range of other cavities, including many designs of artificial beehives. Subspecies vary in aggression, with some African races very aggressive. Some races, such as those from southern Africa, swarm readily in times of dearth, and are hence less easy to manage than the European races.

CHAPTER 7

The Challenges Faced by Bees

An Introduction to Bee Losses ❧

In recent years, the world's media have reported that bees seem to be dying at an unprecedented rate. Certainly there have been significant losses of honey bees, but is this a new thing, and are the losses restricted to honey bees?

HISTORIC BEE LOSSES

Calamitous losses of the western honey bee (*Apis mellifera*) have occurred at various times throughout history and throughout the world. For example, in the USA, significant losses were reported in Kentucky in 1868, then in many places in the early twentieth century, and again in the 1960s and 1970s.

More than a hundred years ago, substantial losses of bees in the United Kingdom were attributed to the "Isle of Wight disease." This was first seen on the Isle of Wight, an island off the south coast of England, and some claimed that it nearly wiped out the native British black honey bee (*A. mellifera mellifera*), although the bees remained common elsewhere in Europe.

The leading entomologists of the early twentieth century worked to uncover what lay behind these bee losses, and in 1921 the cause was identified as acarine, or tracheal, mite (*Acarapsis woodi*). The

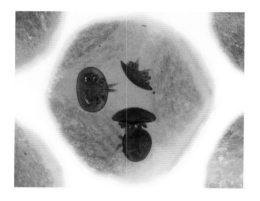

Left *The* Varroa *mite is the most destructive parasite of honey bees worldwide. Here, the mites have been lured to one cell of the comb that has been treated with an attractant chemical—part of a research project to find ways of controlling the pest.*

symptoms caused by this mite, however, do not coincide with the bees' symptoms as reported at the time. During the 1950s, Leslie Bailey at Rothamsted Experimental Station concluded that the Isle of Wight disease was almost certainly caused by chronic bee paralysis virus, a previously unknown infectious disease.

RECENT LOSSES

The losses that have most hit the news are those of commercial honey bees, especially those used for pollination of almonds in California and apples in Pennsylvania. In 2006 many bee farmers returned to their hives to discover most of the bees were gone. It was at this point that the term colony collapse disorder (CCD) was first used.

Although they rarely appear in the headlines, many of the world's other bee species have also been in decline over the last fifty years, and particularly over the last twenty. As these bees live in the wild, and are not kept in hives, their disappearance has been less obvious—and they have been in decline for far longer than just the last few years. For example, the UK has lost three bumble bee species in the last 150 years. The apple humble bee (*Bombus pomorum*) has not been seen since 1864; Cullum's bumble bee (*B. cullumanus*) was last recorded in 1941; and the short-haired bumble bee (*B. subterraneus*) vanished in 1988. However, all these species are found, sometimes commonly, outside of the UK.

The term bumble bee scarcity syndrome was coined in Europe in 2012 to describe situations where the number of bumble bees seen was abnormally lower than expected due to heat waves. These losses extend to the Americas and Asia. A number of unique North American bumble bee species are declining, with their ranges contracting, and four are scarce: the western bumble bee (*B. occidentalis*), the American bumble bee (*B. pensylvanicus*), the rusty-patched bumble bee (*B. affinis*), and the yellow-banded bumble bee (*B. terricola*). One species— Franklin's bumble bee (*B. franklini*)—is believed to be extinct. In stark contrast, four other North American bumble bee species remain abundant and widespread, including the commercially important common eastern bumble bee (*B. impatiens*) as well as *B. bifarius*, *B. vosnesenskii*, and *B. bimaculatus*.

Worldwide, far less is known about the situation of solitary bees. Initial studies of North American solitary bees indicate that these species are declining at a slower rate of about 15 percent, compared with 30 percent for bumble bees, since the 1870s. A number of Hawaiian species of yellow-faced bees (*Hylaeus* spp.) are critically endangered, and some are potentially extinct.

Right Varroa *mites, like the one on the back of this honey bee (*Apis mellifera)*, pose a serious threat to the health of bees and beekeeping, and may be one of the factors underlying colony collapse disorder.*

Weather & Climate

As a potential cause for bee losses, the weather can have serious local impacts, but such a factor is outside human control. However, global climate change—which could lead to a myriad of changes affecting bee survival—may be within human influence.

THE IMPACT OF WEATHER ON BEES

Since the start of the twenty-first century, winter weather in many countries has become more variable. This pattern seems to tie in with the pattern of honey bee losses in the USA, the UK, and Europe. For example, the US winter of 2011/12 was unusually mild, and it is significant that the loss of honey bee colonies was "only" 22.5 percent nationally, compared with 30.6 percent in the winter of 2012/13 when conditions were severe. Inclement weather has always been harmful to hive bees, and it is now recognized as an important factor in colony collapse disorder.

The negative impact of inclement weather on bee survival may be reversible as the weather becomes more favorable, but if a species is living at the edge of its climatic range or is rare, a run of several years of bad weather can be enough to lead to its local extinction.

CLIMATE CHANGE & BEES

Bumble bees, with their furry bodies, evolved in the cooler regions of the world. Within their normal habitat ranges, increasing temperatures—resulting in both warmer winters and hotter summers—could affect their ability to survive. One feature of climate change is erratic weather patterns. For example, during a day of unseasonably warm weather in midwinter, a queen might emerge from hibernation and fail to find food because it is too early in the year for the right flowers to bloom. Even if she has sufficient energy to return to her overwintering site, her reserves are depleted and she is less likely to survive to the spring.

Recent studies have shown that warm spring weather now starts earlier each year in temperate regions of Europe and North America. Concerns exist that if the bees emerge earlier than the flowers, neither will the flowers be pollinated nor will the bees find food. Few studies have provided evidence for these potential mismatches, however. For a few North American bee species, it was found that over 150 years they now emerge ten days earlier, and so do the flowers they visit. There is a possibility that both plants and

Above *The Asian hornet (*Vespa velutina*), shown here attacking a bee in France, presents a far deadlier threat than the European hornet (*V. crabro*).*

bees might slowly adjust to long-term changes in the climate, but this is by no means certain.

In one study, the distribution of *Colletes* bees was compared with climate-based patterns of expected distribution across Europe. The ivy bee (*C. hederae*) has increased its range rapidly over the last twelve years, but it has not yet filled the range where the climate is already suitable in large parts of Italy, France, and Spain. It visits solely European ivy (*Hedera helix*), which is very climate-sensitive and flowers only in the warmer areas of its range. As the climate warms this ivy is likely to flower farther north, enabling *C. hederae* to expand its range. It is expected that many plants and bee species will shift northward and to higher altitudes as the climate warms, disrupting the existing ecosystems.

Distribution models suggest that generalist bees will adapt better to climate change than the specialist species that are restricted to foraging on a single plant species or genus.

We know little about the direct effects of climate change on plants, bees, and their complex interrelationships. Higher temperatures affect floral scent, nectar, and pollen production in plants, changing their attractiveness to bees in terms of the lure (the scent) and the quality and abundance of the rewards (the nectar and pollen). Furthermore, an inverse relationship exists between temperature and the size of bees, with higher temperatures resulting in smaller bees. If bees become smaller through climate change, then, given that bigger insects carry more pollen and are able to forage over greater distances, the delicate balance between pollinator and plant could easily change.

Habitat Loss

Land-use changes are harmful to bees when habitats that were once suitable for them are lost or damaged. Once commonplace species have now become regionally extinct, and they may only exist at low levels elsewhere.

LOSS OF WILD HABITATS

Much land has been "reclaimed" for human use, often for farmland. Between 2006 and 2011, more than 500,000 hectares of grassland was reclaimed from the shallow wetlands of the Great Plains prairie pothole region of North America, including North Dakota, with an 80 percent loss of habitat for wildlife species. The impact of agriculture on the range and number of native bees, potential bee nesting sites, and suitable flowering plants is being assessed. Similar studies are also being carried out on non-native honey bees—because North Dakota produces more honey than any other US state (32.7 million tonnes in 2012).

Australia has a diverse and unique native bee population. Land clearance and agriculture, resulting in the loss of forage and nesting sites, has adversely affected many Australian bees. For example, grazing has severely reduced western myall trees in the southern arid regions, with the loss of nest sites used by the allodapine bee *Exoneurella tridentata*. Steps to conserve native Australian bees have been hampered by a lack of taxonomic expertise and suitable identification keys. In 2009, some 300–400 species remained undescribed.

URBANIZATION

Half of the world's human population lives in towns and cities, and predicted population growth is likely to create ever greater tension between the competing demands for housing, agriculture, and preservation of the natural environment. Urbanization typically decreases biodiversity as it encroaches on rural habitats, and the remaining rural habitats are degraded through agricultural intensification.

Urbanization transforms the landscape, bringing habitat disturbance and loss. It eliminates nest sites and native forage, leading to fewer bee species.

Above *Plants grown in urban environments, such as this chokeberry or Siberian blueberry (Aronia sp.) in a pot on a balcony, can provide a valuable resource for local bees.*

Of particular concern to ecologists is habitat fragmentation. Pockets or islands of suitable habitat are surrounded by inhospitable areas. These isolated habitat fragments are less likely to retain their native insect populations. Part of the solution lies in optimizing green spaces for bees. Bees can benefit from ribbons of pollinator-friendly habitat linking fragments of good habitat through reserves, parkland, and gardens.

Honey bees are able to gather supplies from a wide range of plants, so they thrive in cities, provided the number of colonies does not exceed what the local floral resources will support. Native bees, however, do not always fare so well. In the Polish city of Poznan, 104 native bee species were collected over three years—but one-third of the individuals belonged to just five species. Almost 90 percent of the individual bees (and three-quarters of the species) were generalists, with specialist bees accounting for less than 1 percent of the bees seen.

Below *Urban beekeeping: John Chapple of the London Beekeepers' Association tends an urban rooftop hive in London, UK.*

PLANTING FOR BEES

Urban areas often have a wider range of flowering plants than natural sites—and certainly more than intensively farmed fields—but comparatively little is known about the impact on the diversity and composition of bee populations. When the number of plant species was doubled in one Californian community garden, the number of native bee species rose from five to forty. The plants flowered successively from early spring to late fall, thus extending the bees' foraging season. The project was shared with the public through education and outreach programs, thus spreading the words about helping bees through gardening.

Green roofs may have a significant role to play in bee conservation, and the number of projects across the world grew from 93 in 2000 to over 1500 in 2013. In Toronto, Canada, bee communities on green roofs were very similar to those at ground level, confirming the results of earlier studies in London, UK and Basel, Switzerland.

Changes in Agriculture 🐝

Many wild bee declines have been driven by changes in agricultural policy and practices in Europe, North America, and China, especially in the last fifty years. Bees are particularly sensitive to agricultural intensification. To survive in a modern agricultural landscape—or indeed any landscape—bees need nest sites and sufficient food plants for nectar and pollen. Abundance and diversity of bees and other pollinators is often far lower on a crop than on the surrounding vegetation.

TRADITIONAL CROP ROTATION

During the reign of Charlemagne in the early ninth century, a three-year rotation was widely adopted throughout Europe. A green manure—usually a legume (peas, beans, or lentils)—was grown on one-third of arable land every year. This crop was used to fix nitrogen in the soil to increase its fertility. Legumes are excellent forage plant for bees, especially for long-tongued bumblebees. Then the land was left unsown (fallow) for one year, which allowed annual wild flowers to flourish, increasing the food resource for bees. In the third year a cereal, usually wheat or rye, was grown. Farming essentially did not change in Europe until the eighteenth century, when a four-year rotation was widely introduced. The fallow year was eliminated and legumes were grown only one year in four. This agricultural revolution reduced the bee

Left Engraving of farm workers bringing in the harvest during the medieval period.

Right *Red clover (*Trifolium pratense*) was used to enrich meadows for livestock before the advent of chemical fertilizers, and is a key food plant for bees, like this bumble bee (*Bombus pascuorum*).*

forage from up to two-thirds to only one-quarter of arable land. Sadly, we will never know the impact this change had on bees, because we have no records.

THE IMPACT OF FERTILIZERS

Early in the twentieth century, the development of "chemical" fertilizers made redundant both the arable crop rotation and the use of clover. Yet more bee forage was removed from the landscape as improved mechanization made it easier to apply these chemicals. Farmers became able to grow the same crop year after year.

Red clover (*Trifolium pratense*) remains an important forage crop in Sweden. Seed set is dependent on pollination by bumble bees. However, among other changes in the management of farmland, there has been a shift from traditional hay meadows and clover-rich pasture to chemically fertilized grass-rich fields that support a higher density of livestock, but a much lower density of bees.

Over the last seventy years the bumble bee population has altered. The long-tongued bumble bees that most effectively pollinate red clover (*Bombus pascuorum, B. hortorum, B. sylvarum,* and *B. distinguendus*) have either become scarce or endangered in Sweden and Denmark. In marked contrast, short-tongued bumble bees (*B. terrestris* and *B. lapidarius*) have become dominant as they can exploit other flowers that are available.

In New Zealand, it was found that the yield of red clover declined through shortage of bumble bees when fields exceeded 5 hectares. In large fields, there is proportionately less field margin and hedgerow to provide nesting habitats for bees and other pollinators.

Modern Farming

MECHANIZATION

The advent of farm mechanization reshaped the farming environment to one that suits bees less well. From a grueling and largely manual occupation, it became possible to farm enormous tracts of land with minimal labor. Henry Ford's Fordson tractors were first produced in 1917. Modern plows dug more deeply into the soil, harmful where soils were light. In combination with severe drought and ever larger fields, the North American Great Plains and prairies were turned into arid dustbowls in the 1930s. As the native grasses and flora were no longer there to secure the soil, tonnes of topsoil were blown across the prairies. The seeds of native plants that would have fed the native insect pollinators were also lost.

In Europe, too, mechanization has transformed the landscape. Even an apparently simple change such as switching from making hay in August to cutting grass for silage in June or July has caused the destruction of nests of the surface-nesting bumble bee *Bombus pascuorum*. The mowing process destroys newly established nests, killing some queens and forcing others to start again. Intensive crop and grassland management, both plowing and cutting, has led to bumble bee nests being concentrated in the less disturbed field margins, hedgerows, and unmanaged strips of land.

Left *Early farm mechanization: a Fordson tractor and other machinery in 1922.*

CONVENTIONAL FARMING

Notably since the 1940s, farmers have planted extensive monocultures, placing heavy reliance on fertilizers and agrochemicals. Moreover, as the size of the machinery has increased so has the size of fields. The aim is to make farming more efficient and productive, but the cost is measured in the loss of wetlands, woodlands, and meadows. Hedgerows have been replaced with wire fencing, and field margins have become smaller as more land is plowed.

Such changes have resulted in the elimination of the ribbons of land that were a favored habitat for pollinators, compounded by the application of low-cost fertilizers, making the soil too rich for many traditional bee forage plants. Modern herbicides have also killed off many flowering weeds that previously would have grown within the crop, at field margins, and on roadside verges.

Below *The modern face of farming, exemplified by large machinery in an open field: crop irrigation using a center-pivot sprinkler system.*

IS ORGANIC FARMING GOOD FOR BEES?

By switching from conventional to organic farming, it has been calculated that a farmer could increase the abundance of wild bees by 74 percent and their diversity by 50 percent.

Organic farming results in biodiversity gains, but even if more farmers switch to organic methods the good news may not last. Because the crop yield is lower, organically managed farms may become large monocultures—and this is predicted ultimately to be detrimental for wild bees and their pollination services.

Much of the wildlife gain on organic farms is because they tend to be small and grow mixed crops. Use of managed field margins, hedgerows, and nature reserves might be more effective ways to promote biodiversity.

PROMOTING BIODIVERSITY

Wild bees alone can fully pollinate crops in certain circumstances, and, even when honey bees are present, wild bees can enhance rates of pollination. Their diversity and abundance is influenced by local farm management and the quality and structure of the local landscape.

In agricultural landscapes, crops are typically grown in extensive monocultures even where they are pollinated by bees. Often these crops—such as almonds, cranberries, and canola—provide forage only for a short period, and once the crop has flowered they offer nothing for the bees. Even if there is little good-quality natural habitat, wild bee populations can be supported by using organic farming methods, reintroducing crop rotations, or adding hedgerows and field margins offering good bee habitats—always provided there remain sufficient local populations of wild bees.

Other approaches include planting small fields with mass-flowering plants and managing low-input meadows or semi-natural woodlands. Just a 10 percent increase in suitable (semi-natural) bee habitats within a landscape can increase the diversity and number of wild bees by 37 percent.

Given that over one-third of the total land area of the USA (some 360 million hectares) is farmed, the management of farmland can impact bees significantly. Programs such as the US Farm Bill, for example, pay farmers to set aside land to be used as wildlife habitats. Other greening projects exist in other countries, such as the European Union's Set-aside project, introduced in 1992 as part of the Common Agricultural Policy (CAP), and more recent CAP reforms which provide payments for farmers if they satisfy certain environmentally beneficial conditions.

Left Cranberries almost ready for harvesting—a monoculture that provides rich forage for bees, but only for a short period.

GENETICALLY MODIFIED CROPS

Among many concerns, fear exists that genetically modified (GM) crops might have insecticidal properties that harm bees. That concern has not been verified; when bees were fed with pollen from the GM crop Bt-corn, it affected neither bee gut microflora nor the development of the hypopharyngeal gland. The toxin Cry1Ab produced in Bt-corn did not affect western honey bee learning.

However, farmers use herbicides on GM herbicide-resistant crops to kill weeds within the crop and at the field margins. This could eliminate key floral resources, damaging the local ecosystem and severely affecting pollinator populations. An outright ban on GM crops exists throughout the European Union, but many other countries allow them. Across the world, there is considerable variation in regulations affecting whether GM crops may be grown or their products used.

Agrochemicals ❧

INSECTICIDES

Farm crops are attacked by many pests and diseases, and pesticides have been developed to protect them from attack. There is inevitably a risk, however, that an insecticide intended to kill pests might also be harmful to a range of beneficial insects such as bees.

Concerns about pesticides first became apparent in the 1940s, when fruit trees were protected with sulfuric acid, arsenic, and heavy metal compounds. The next decade saw the widespread use of the first generation of synthetic insecticides, the organochlorines—including DDT. It soon became apparent that these had unforeseen side effects on non-target

Right *Alfalfa fields in Imperial Valley, California, used to be sprayed with DDT.*

animals such as birds of prey, leading to the pioneering work of Rachel Carson. In her 1962 book *Silent Spring* she drew attention to the complex food chains and webs that connect very different species, meaning that pesticides may affect organisms seemingly unconnected to the target. Her work directly led to the start of the modern environmental movement.

The next generation of insecticides, the organophosphates, proved to be very damaging to bee populations when used on flowering crops such as oilseed rape. In the UK, thorough records of honey bee pesticide poisoning incidents have been kept, and these show many incidents involving the death of hundreds of colonies in the 1970s and 1980s due to organophosphates.

The introduction of safer compounds, the carbamates and then the synthetic pyrethroids, often marketed as "bee-friendly," led to a dramatic reduction in poisoning incidents. For many years there was no confirmed incident of poisoning of a honey bee colony in the UK that could be linked to the approved use of an agricultural pesticide.

In the mid-1990s, however, a new class of insecticide was introduced. These were the neonicotinoids. Although very toxic to bees in the laboratory, they were intended to be used as a systemic seed dressing, the idea being that the chemical would be taken up by the growing plant, protecting it from sucking and biting insects. In theory, this should have been safer for bees, as they would no longer be exposed to insecticide sprays. But very soon after these compounds, in particular imidacloprid, began to be used on sunflowers in France, beekeepers noticed sudden losses of honey bee colonies, resembling the damage caused by the *Varroa* mite. Campaigning by beekeepers led to restrictions on the use of imidacloprid on sunflowers and maize (corn), but losses continued. In hindsight, these early losses in France are not likely to have been caused by the pesticide, but were probably due to inadequate control of *Varroa*. Coincidentally, populations of the mite which had become resistant to the chemicals used to control it appeared in France at exactly this time. Subsequently, large-scale field studies carried out over a number of years failed to link damage to honey bees to the use of neonicotinoid insecticides.

In 2008, however, sudden losses of many honey bee colonies occurred in the Rhine Valley, Germany, and in neighboring Italy and Slovenia, with bees exhibiting the classic symptoms of acute pesticide toxicity. These incidents were rapidly linked to the use of another neonicotinoid, clothianidin, as a seed dressing on maize. It was discovered that the chemical was being used at a high rate, and factories dressing the seed had failed to use a vital ingredient to make

Below *Rachel Carson, author of the book that gave birth to the modern environmental movement.*

Right *A protest against neonicotinoids in front of the European Parliament in Brussels, Belgium.*

the dressing "stick" to the seed. The seeds were being planted using pneumatic seed drills, and this led to the production of highly toxic dust that was blown onto neighboring oilseed rape on which bees were foraging.

Recent laboratory studies have shown that neonicotinoid insecticides may also have more subtle, non-lethal effects on bees. Bees fed syrup laced with insecticide have been shown to have shortened lives and impaired memory, homing, and foraging behavior. Most convincingly, a study on bumble bees showed when bees were fed with doses of imidacloprid the dosed colonies were slightly smaller than control colonies that had not been given any of the chemical. Crucially, the imidacloprid-fed colonies produced only a fraction of the new males and queens necessary to ensure colonies the following year. So far these effects have not been observed in the field, but the European Union was sufficiently concerned to implement a two-year moratorium on the use of three neonicotinoids on bee-friendly crops from 1 December 2013. It remains to be seen what effects this will have on bees and farming.

HERBICIDES & MIXTURES OF AGROCHEMICALS

Herbicides became available in the 1940s. They have killed many weeds, both within the crop and at the field margin, that might have provided good bee forage. It is worth noting that tank mixes, where a cocktail of agrochemicals are mixed and applied together, have long been an issue for beekeepers. In 2003, studies verified this fear, showing that some fungicides significantly increased the toxicity of an insecticide, lambdacyhalothrin, while the fungicide chlorothalonil decreased the repellency of what would otherwise have been relatively innocuous doses of pyrethroid insecticide.

Pests & Diseases

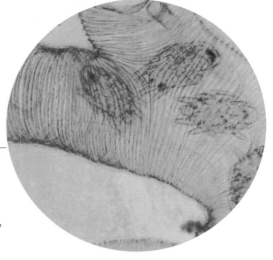

Humans have helped to spread the western honey bee far and wide across the world, taking with it pests and diseases such as American foulbrood, and en route it has collected new afflictions, including the *Varroa* mite and the Asian honey bee version of *Nosema* (*N. ceranae*).

COLONY COLLAPSE DISORDER (CCD)

A variety of pests and pathogens can cause problems for the western honey bee. One of the biggest concerns in recent years has been the condition known as colony collapse disorder (CCD). A colony loss due to CCD is characterized by a rapid loss of adult worker bees from an affected colony, with an unusually large amount of brood for the size of the adult bee population and few or no dead worker bees in or near the hive. The loss cannot be attributed to another pest such as small hive beetle, or to robbing by another hive—although these pests may have appeared after the colony has begun to collapse—and the colony is not necessarily suffering from damaging levels of *Varroa* or *Nosema*. CCD defined thus is a condition seemingly unique to the USA. This does not mean there have not been significant losses of honey bee colonies elsewhere—it just means the strict criteria of CCD have not been met.

PESTS & DISEASES CAUSING COLONY LOSSES

Many studies have attempted to identify the causes of honey bee losses. One report stated that CCD was closely linked with four ailments: *Nosema apis*, *N. ceranae*, Kashmir bee virus, and Israeli acute paralysis virus. Yet the *Varroa* mite is generally agreed to be the most significant cause of colony losses, because of the viruses it transmits.

Maps of worldwide western honey bee losses indicate losses arise wherever *Varroa* is found—but not, for example, in Australia, where *Varroa* is not found. *Varroa* mites are present in South America and Africa, yet losses there are low. This is probably because African honey bees and the Africanized honey bees in America have an innate tolerance to *Varroa*. A number of factors are involved, including a shorter brood development time in Africanized bees (so fewer mites complete their reproductive cycle) and less virulent strains of *Varroa* occurring in those places.

Above *A micrograph showing a bee trachea infested with mites.*

PESTS & DISEASES
OF OTHER BEES

Various other pests and diseases affect non-*Apis* bees, but they are generally less well understood, if studied at all. Some studies have been made of a number of bumble bee species, however, because they are important pollinators.

A bumble bee of great commercial importance is the buff-tailed bumble bee (*Bombus terrestris*), which is used to pollinate a range of greenhouse crops. Like the western honey bee, it has been transported across the world to many areas that are outside its natural range. One such place is Japan, where *B. terrestris* has escaped and become established in the wild. Meanwhile, a Japanese bumble bee (*B. ignitus*) was also taken back to Europe for breeding and was later returned to its native Japan. Both groups of imported bees, *B. terrestris* and *B. ignitus*, were found to be infested with a European strain of a bumble bee tracheal mite (*Locustacarus buchneri*). This parasite has since infested wild populations of the Japanese native bumble bees, and their populations are now declining.

Non-native strains of *Nosema bombi* have been implicated in the rapid decline of US bumble bee species. Recently, a number of infectious bee pathogens have been found in "parasite-free" commercially produced bumble bee colonies, despite this practice being heavily regulated to prevent parasites being spread.

Below *US Department of Agriculture scientists are investigating the possibility that transfer of food or other fluids among members of a bee community through mouth-to-mouth feeding might act as a facilitator of colony collapse disorder.*

Other Threats to Bees

QUEEN FAILURE

Over recent years, many beekeepers have observed that western honey bee queens fail to produce viable eggs for one whole season, even though in the past they were effective for three or more years. There could be several reasons for this, but Jeff Pettis of the US Department of Agriculture has recently worked out that some of the chemicals used by beekeepers to control *Varroa* persist long after the initial application, and affect the viability of the sperm stored by the queen. If a large proportion of the sperm die, the queen will tend to become a drone layer and will be superseded within the hive. With the chemical persisting in the hive, it can be seen that a succession of queens might fail, leading to colony loss.

OTHER REASONS FOR COLONY FAILURE

A survey carried out over the winter of 2012/13 showed that the main reasons for colony failure were shared by all types of US beekeepers. Colony losses were no greater for the beekeepers who migrated their bees across the states, including the colonies that pollinated the Californian almond orchards. Colony collapse disorder was reported only by the commercial beekeepers, but it is noteworthy that many backyard beekeepers did not know why their bees had died (reminding us that the reliability of data such as these depends on the ability of beekeepers to diagnose problems correctly).

Other key reasons given by US beekeepers for colony loss included weak colonies in the fall and starvation. Both conditions are avoidable through good colony management, and issues such as these should lead responsible beekeepers to question whether they are taking appropriate and timely actions to care for their bees.

In contrast, South African data suggests that the major cause of colony loss there is due to the Cape honey bee, which behaves as a social parasite, destroying local *Apis mellifera scutellata* colonies.

Left *Investigating the causes of colony collapse disorder: entomologist Jeff Pettis inspects honey bee combs at Beltsville, Maryland, for disease.*

Right In Tasmania, pollination by escaped buff-tailed bumble bees (Bombus terrestris) has enabled the introduced invasive weed purple loosestrife (Lythrum salicaria) to flourish, supplanting much of the native flora.

ENVIRONMENTAL DAMAGE FROM IMPORTED BEES

Importing bees may also affect the environment. The western honey bee (*A. mellifera*) and the buff-tailed bumble bee (*Bombus terrestris*) are both "generalists." As long-tongued bees they can forage on a wide range of flowers, potentially outcompeting the native bees at gathering supplies. Typically, they forage earlier in the day and deplete supplies of nectar and pollen that would otherwise have been available for native bees.

Imported bees are often used to "saturation pollinate" an area, so affecting all plants in the area. The western honey bee is as effective as the native bees at pollinating the Australian plant *Banksia ornata*. But if imported bees preferentially pollinate any introduced plant species (as well as the intended crop), but not the native wild flora, then the ecosystem will change as the native plants lessen and the foreign weeds proliferate—as seen with purple loosestrife (*Lythrum salicaria*) in Tasmania.

Finally, many regions have local bee strains that are well adapted to local conditions. For example, the long-furred *audax* strain of *B. terrestris* thrives in the cool damp UK climate—and the longer fur also holds more pollen, making it a better pollinator. Yet the short-furred southeast European *dalmatinus* strain has been used for greenhouse pollination in the UK. It has escaped and hybridized with local *audax* bees. The offspring have thrived, weakening the local populations of the species.

USING THE HONEY BEE AS A MODEL TO UNDERSTAND BEE AND POLLINATOR LOSSES

It is not just bees that are under threat. Many other pollinators, including bats, flies, and butterflies, are also in decline. Moreover, there are thousands of bee species, and very little is known about most of them, so it is difficult to understand or predict how environmental or other factors may influence their survival. Ecologists have suggested that the western honey bee, given the high level of reliable data collated worldwide, could be used as a model insect pollinator in a wider study of pollinator declines and pollination under continuing environmental change. A multidisciplinary approach, combining knowledge of the species' biology with a study of aspects such as land use, could help us to understand and find ways to remedy these pollinator declines.

Fear of Bees

With all the other threats they face, bees really need us on their side. The media have widely reported the recent losses of bees, leading to a range of responses from interest and concern to outrage, but they also often report that bees sting and sometimes kill humans. Yet bees, unlike wasps, are likely to sting only if their colony is threatened or they are being squashed. As we have seen, humans are not always friendly towards bees, but we may be less likely to want to help them if we believe that they are the enemy.

A CASE OF MISTAKEN IDENTITY?

Most people's fears of "bees" are fueled by a lack of knowledge or by mistaken identity. In the United States, the yellow jacket (*Vespula* spp.) and the paper wasp (*Polistes* spp.) can sting repeatedly with their smooth (unbarbed) sting. These are aggressive social wasps nesting underground, in hollow trees, in crevices, or under the eaves of houses and similar structures. These wasps have a thin waist, lack pollen baskets, and are less hairy than bees, but many people still mistake them for bees. Similar black and yellow wasps are found throughout Europe and Asia, including the common wasp (*V. vulgaris*), the German wasp (*V. germanica*), the median wasp (*Dolichovespula media*), and the European paper wasp (*P. dominulus*).

AFRICANIZED HONEY BEES

One bee found in the Americas that merits caution, and causes much fear, is the Africanized honey bee. It resembles a western honey bee, and is in fact a hybrid between one of the western honey bee strains and the southern African subspecies *Apis mellifera scutellata*. On average, two deaths due to Africanized honey bees are reported each year in the USA. Usually, the victims were allergic to bee stings. These bees were first found in California in 1985 and in Texas in 1990, and it is fortunate that their rate of spread is slow. It likely their range will be limited as they encounter more temperate zones.

Left *A yellow jacket (*Vespula *spp.).*

Above *An Africanized honey bee gathering pollen from a* Cosmos *flower.*

PHOBIAS & SOCIAL TENSIONS

A fear of bees can interfere with an enjoyment of being outdoors, and it can develop to become a medical condition, apiphobia. Fear is understandable in someone who has been stung, and it can be magnified if that person has an allergy to the venom. But bee phobia can also arise from knowing someone who has been stung or who is allergic to stings.

It is worth noting that the venoms of different species differ. Should an individual be allergic to wasp venom then that person is not automatically allergic to bee venom (and vice versa).

The western honey bee was imported to Brazil in 1839, but the climate proved unsuitable for the European strains. In 1956, a Brazilian scientist imported bees from South Africa, suggesting that they would prove more suitable for the climate. Twenty-six colonies of these bees escaped, and queens and drones mated with the "local" western honey bees to produce hybrid bees, which were well suited to the climate but had a number of undesirable traits. Africanized honey bees swarm and abscond frequently, readily forming feral colonies, and are very aggressive, attacking in response to only minimal disturbance. They are unusually persistent, and a mass of bees can pursue the victim a considerable distance from the original attack.

Africanized honey bees are, however, excellent honey makers, and Brazil is now the world's eleventh-largest producer of honey. These bees also make much propolis, used in the pharmaceutical industry, giving the beekeepers an additional income stream. Although keeping these bees in a backyard is impossible, Brazilian beekeepers have learned to manage them by wearing protective suits and by using giant smokers to calm the bees, gentle handling, and selection of gentler stock. These bees are much liked by commercial beekeepers because they are very productive and highly resistant to disease, while the risk of theft is low.

Pollinators, the Environment & Conservation 🦋

In temperate zones, 78 percent of flowering plants rely on animal pollinators, and in the tropics the equivalent figure is 94 percent. In the tropics many plants rely on birds and bats as pollinators, as well as bees, moths, and butterflies. Without pollinators, we would stand to lose a wide range of plants, all of which have a vital role in biodiversity and ecosystems.

THE NEED TO ASSESS OUR WILD BEES

One concern is that a bee species could die out, yet its loss might not be noticed until a long-lived perennial plant for which that bee species was the only pollinator suddenly disappeared from our planet forever. With a few exceptions, such as in the UK and the USA, we lack the systematic studies that would enable us to ascertain the health of wild bee populations.

In the Mediterranean region, for example, we have woefully inadequate data on the rich and important bee populations. In 2012, a three-year program commenced to assess the status of native insect pollinators of mainland Greece, Turkey, and the Aegean islands. Other countries have started or will soon start similar studies.

Below *Beehives on the Mediterranean island of Crete, Greece.*

CONSERVING RARE BUMBLE BEES IN BRITAIN

Many bumble bee species that were once widespread throughout the United Kingdom have now retreated to the Scottish coast and islands. These environments are managed less intensively and typically include nature reserves—where, in the absence of farming, fertilizers, and pesticides, wild native flora remains plentiful.

The great yellow bumble bee (*Bombus distinguendus*) is one formerly common species that is now endangered, largely due to intensive farming since the 1960s. Now restricted to the machair, a unique wild grassland habitat found on Hebridean islands and the coastal grasslands of north and west Scotland, the bee is slowly recovering, aided by the restoration of natural habitats and the planting of borage, a flower upon which it forages. These practical measures were undertaken by a partnership of conservation organizations using teams of volunteers who planted suitable habitats, made observations, created public awareness, and worked with farmers and crofters to help this rare species.

DEDICATED BEE RESERVES

When honey bee populations in Britain were devastated in 1906 by the Isle of Wight disease, beekeepers imported foreign strains of *Apis mellifera*, especially from Italy (*A. m. ligustica*) and Germany (*A. m. carniola*), which interbred with the native British black bee (*A. m. mellifera*), resulting in bad-tempered and very variable mongrel bees. Some beekeepers have sought to protect the remaining

THE BIRDS & THE BEES

Fruit and berry set on hedgerow bushes such as blackthorn, hawthorn, and ivy is greatly reduced if insect pollinators are excluded, and yet these fruits are important winter and spring food for birds. If the fruits were no longer produced, then it is reasonable to conclude that the wild birds that feed on them would also decline. Moreover, these bushes also supply shelter and nesting sites, so in the longer term the loss of this vegetation would be very harmful to the birds.

pure lines of the British black bee. In January 2014 a protective reserve was established on two islands in the Hebrides under new legislation. It is now illegal to keep any type of honey bee other than the black bee on these islands. The only other examples of this type of reserve exist in the Swiss canton of Glarus, and in Tasmania, Australia, although the Tasmanian reserve has been badly affected by deforestation and the introduction of feral Italian honey bees.

Research Initiatives to Help Bees 🐝

UNDERSTANDING THE DECLINE IN BEES & OTHER POLLINATORS

In 1996, the Convention on Biological Diversity directed that international research efforts were necessary to understand pollinator losses. The International Pollinator Initiative was launched in 2000 with the creation of various pollinator partnerships worldwide, including the North American Pollinator Protection Campaign. These groups have set up and coordinated the efforts of scientists, public servants, industry, farmers, and not-for-profit organizations to explore the full scope of the pollination problem.

In 2006, news of widespread and alarming losses of the western honey bee led to further research. In the USA, for instance, the Bee Informed Partnership started gathering survey data, and following the severe losses seen in the winter of 2012/13 the partnership launched an emergency response kit to identify the causes of colony losses for beekeepers whose colonies are failing.

COLOSS (Prevention of honey bee COlony LOSSes) started as an EU project and evolved into a network of more than three hundred researchers from over sixty countries who seek to better understand honey bee health and prevent further heavy losses of colonies worldwide. In 2013 COLOSS published the *BEEBOOK*, which, for the first time, sets out standard protocols for honey bee research, giving scientists common techniques so that the results of future bee research will be directly comparable, helping us to better understand bee health and prevent further losses.

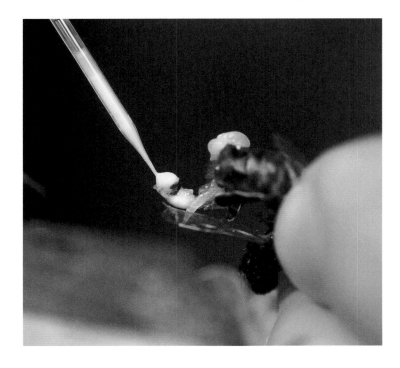

Below *Bee research: collecting semen from a drone honey bee (*Apis mellifera*) for use in artificial insemination.*

naturally show variable levels of hygienic behavior. By artificially selecting for such traits, the Minnesota team have bred bees that show increased resistance to multiple pests and pathogens, including the *Varroa* mite. If the bees remove infested brood, then fewer mites mature to infest the colony. Further work on breeding better bees is also being undertaken in various countries worldwide by members of COLOSS. Research in Argentina and Poland has shown that desired traits become apparent after about four generations of selective breeding.

PROBIOTICS FOR HONEY BEES

In 2005, two scientists in Sweden discovered unique strains of *Lactobacillus* spp., *Bifidobacterium* spp., and other symbiotic bacteria in the honey crop of western honey bees, and, at the same time as some American scientists, found that these bacteria protect bees from foulbrood diseases. One result of this research has been the development of probiotic supplements for honey bees, designed to fight disease.

The newly isolated forms of lactic acid bacteria have been found in stingless bees as well as in a range of honey bee species, suggesting they have co-existed with bees throughout their evolution. Brood food contains these bacteria, which, when fed to developing bees, enhances the young bees' innate immune system. Although the value of probiotics to bees in the field remains to be tested, what is clear already is that the use of antibiotics to combat American foulbrood may undermine this important relationship between the beneficial bacteria and the bee.

Above *Understanding the declines in bee populations necessitates detailed research on every aspect of their biology. Here, US Department of Agriculture scientists look for signs of disease.*

BETTER HONEY BEES

Since 1993, Dr. Marla Spivak and her research team at the University of Minnesota, have achieved impressive results with a selective breeding program to produce "hygienic bees" that remove unhealthy and dead brood from the nest, thereby eliminating potential pathogens from the hive. Honey bee colonies

How We Can Help the Bees

One obvious way of helping honey bees is to become a beekeeper. If you are not keen on becoming a beekeeper yourself, you could consider allowing bee hives to be placed on your land—though this isn't for everyone, and local laws may even prevent the keeping of honey bees in some areas. But beekeeping is not the only answer—there are many other things you can do to help bees.

MAKING A HOME FOR BEES

Even a window box can be a substitute for lost habitats. Plants with single and open flowers are a good choice. You should avoid the highly bred double and fancy flowers, which may produce nothing for the bees, or in which the nectar and pollen are out of their reach.

Traditional cottage flowers and native plants are often highly attractive and garden-worthy. Seed bombs, made by combining local native flower seeds with soil, can be used to prime suitable sites with good forage. It is important to provide a range of flower types to suit different pollinator species, and important that they flower in succession to provide a continuity of nectar and pollen.

Lists of plants beneficial to bees are widely available. Reliable sources include the Royal Horticultural Society in the UK, the Xerces Society in the United States, various university education extension sites, beekeeping groups, and local garden centers and nurseries.

Ready-made houses for mason, bumble, and solitary bees can be bought and placed in a quiet spot—but they are also simple to make, as explained in Chapter 5. Something as simple as a pile of undisturbed leaves will potentially provide a home for wild bees.

If you do grow plants, it will help the bees greatly if you grow them organically and do not use insecticides, which inevitably kill bees.

CITIZEN SCIENCE

This is an exciting new area of interest and study. There is an increasing number and range of projects in which ordinary people are encouraged to take part, and their efforts feed into larger schemes to understand our environment and its inhabitants. Good examples are run by the Bumblebee Conservation Trust in the UK and the Xerces Society, who ask their members to send in photos of bumble bees they have seen—but many other projects exist.

Above *Garden flowers such as lavender (*Lavandula *sp.) can provide rich forage for honey bees, bumble bees, and many other insects.*

LOBBYING

The political consensus is that we should do what we can to help bees, yet in times of recession governments seek to minimize costs, so are reluctant to implement effective bee-friendly measures. It is always worth reminding national, regional, and local government that bees are important, and that they will benefit from better regulation and funding for conservation and research projects.

THE FUTURE FOR BEES

Bees have pollinated flowers for millions of years, but increasingly they need help from us as we damage their habitats. Already, there have been some success stories, such as the recent reintroduction of the short-haired bumble bee to the UK. Scientists and conservationists are making progress. Yet everybody can play their part through something as simple as growing some attractive plants in the backyard. Taking a lesson from the bees themselves, and working together, we can make a difference to the future of the world's favorite insect.

ENVIRONMENTAL RESTORATION

Conservation activities can be both worthwhile and fulfilling. Why not consider becoming a volunteer to record bees, plant gardens for bees, or restore habitats? Learn what you can and help to educate people about bees and why they matter. Whole teams of people worked together to make a difference in the case of the great yellow bumble bee. Find out more from the Xerces Society, the Bumblebee Conservation Trust, or a local group.

Useful
Resources

Bibliography

BOOKS

Bailey L., Ball B. V. (1991) *Honey Bee Pathology*. Academic Press, London.

Benton E. (2006) *Bumblebees: the Natural History and Identification of the Species Found in Britain*. Collins, London.

Carreck N. L., ed. (2011) *Varroa: Still a Problem in the 21st Century?* International Bee Research Association, Cardiff.

Crane E. E. (1999) *The World History of Beekeeping and Honey Hunting*. Duckworth, London.

Davis C. F. (2004) *The Honey Bee Inside Out*. Beecraft, Stoneleigh.

Davis C. F. (2007) *The Honey Bee Around and About*. Beecraft, Stoneleigh.

Dietemann V., Ellis J. D., Neumann P., eds. (2013) *The COLOSS BEEBOOK*, Volumes 1 & 2. International Bee Research Association, Cardiff.

Free J. B., Butler C. G. (1959) *Bumblebees*. Collins, London.

Goodman L. J. (2003) *Form and Function in the Honeybee*. International Bee Research Association, Cardiff.

Goulson D. (2010) *Bumblebees: Behaviour, Ecology, and Conservation*, 2nd edition. Oxford University Press, Oxford.

Kirk W. D. J., Howes F. N. (2012) *Plants for Bees*. International Bee Research Association, Cardiff.

Krebs L. (2008) *The Beeman*. Barefoot Books, Cambridge, MA.

Matheson A., Buchmann S. L., O'Toole C., Westrich P., Williams I. H., eds. (1996) *The Conservation of Bees*. Academic Press, London.

Michener C. D. (2000) *The Bees of the World*, 2nd edition. Johns Hopkins University Press, Baltimore, MD.

Morse R. A., Nowogrodzki R. (1990) *Honey Bee Pests, Predators, and Diseases*. Cornell University Press, Ithaca, NY.

O'Toole C., Raw A. (1991) *Bees of the World*. Blandford, London.

Pundyk G. (2008) *The Honey Trail: in Pursuit of Liquid Gold and Vanishing Bees*. Murdoch Books, Sydney, NSW.

Root A. I., Morse R. A., Flottum K. (1990) *The ABC & XYZ of Bee Culture: an Encyclopedia Pertaining to the Scientific and Practical Culture of Honey Bees*. Root, Medina, OH.

Sammataro D., Avitabile A. (1998) *The Beekeeper's Handbook*. Cornell University Press, Ithaca, NY.

Saunders E. H. (1896) *The Hymenoptera Aculeata of the British Islands*. Reeves, London.

Schmid-Hempel P. (1998) *Parasites in Social Insects*. Princeton University Press, Princeton, NJ.

Seeley T. (1985) *Honey Bee Ecology: a Study of Adaptation in Social Life*. Princeton University Press, Princeton, NJ.

Seeley T. (1996) *The Wisdom of the Hive: Social Physiology of Honey Bee Colonies*. Harvard University Press, Cambridge, MA.

Seeley T. (2010) *Honeybee Democracy*. Princeton University Press, Princeton, NJ.

Sladen F. W. L. (1912) *The Humble Bee*. Logaston Press, Woonton.

Vit P., Pedro S. R. M., Roubik D. W., eds. (2013) *Pot-Honey: a Legacy of Stingless Bees*. Springer, New York, NY.

von Frisch K. (1967) *The Dance Language and Orientation of Bees*. Belknap Press, Cambridge, MA.

Winston M. (1987) *The Biology of the Honey Bee*. Harvard University Press, Cambridge, MA.

MAGAZINES & JOURNALS

American Bee Journal.
www.americanbeejournal.com

Bee Culture. www.beeculture.com

Bee World.
www.ibra.org.uk/categories/
bee_world

Journal of Apicultural Research.
www.ibra.org.uk/categories/jar

JOURNAL ARTICLES

Aizen M. A., Garibaldi L. A., Cunningham S. A., Klein A. M. (2009) "How much does agriculture depend on pollinators? Lessons from long-term trends in crop production." *Annals of Botany* 103, 1579–88.

Batley M., Hagendoorn K. (2009) "Diversity and conservation status of native Australian bees." *Apidologie* 40, 347–54.

Carreck N. L., Ratnieks F. L. W. (2013) "Will neonicotinoid moratorium save the bees?" *Research Fortnight* 415, 20–22.

Cresswell J. E., Desneux N., van Engelsdorp D. (2012) "Dietary traces of neonicotinoid pesticides as a cause of population declines in honey bees: an evaluation by Hill's epidemiological criteria." *Pest Management Science* 68, 819–27.

Gallai N., Salles J. M., Settele J., Vaissière B. E. (2009) "Economic valuation of vulnerability of world agriculture confronted with pollinator decline." *Ecological Economics* 68, 810–21.

Genersch E. (2010) "Honey bee pathology: current threats to honey bees and beekeeping." *Applied Microbiology and Biotechnology* 87, 87–97.

Gill R. J., Ramos-Rodriguez O., Raine N. E. (2012) "Combined pesticide exposure severely affects individual- and colony-level traits in bees." *Nature* 491, 105–19.

Graystock P., Yates K., Evison S. E. F., et al. (2013) "The Trojan hives: pollinator pathogens, imported and distributed in bumblebee colonies." *Journal of Applied Ecology* 50, 1207–15.

Grüter C., Menezes C., Imperatriz-Fonseca V. L., Ratnieks F. L. W. (2012) "A morphologically specialized soldier caste improves colony defense in a neotropical eusocial bee." *Proceedings of the National Academy of Sciences* 109, 1182–6.

Heard T. A. (1999) "The role of stingless bees in crop pollination." *Annual Review of Entomology* 44, 183–206.

Henry M., Béguin M., Requier F., et al. (2012) "A common pesticide decreases foraging success and survival in honey bees." *Science* 336, 348–50.

Honeybee Genome Sequencing Consortium (2006) "Insights into social insects from the genome of the honeybee *Apis mellifera.*" *Nature* 443, 931–49.

Kearns C. A., Inouye D. W., Waser N. M. (1998) "Endangered mutualisms: the conservation of plant–pollinator interactions." *Annual Review of Ecology and Systematics* 29, 83–112.

Morse R., Calderone N. (2000) "The value of honey bees as pollinators of U.S. crops in 2000." *Bee Culture* 128, 1–14.

Neumann P., Carreck N. L. (2010) "Honey bee colony losses." *Journal of Apicultural Research* 49, 1–6.

Nicholls C. I., Altieri M. A. (2013) "Plant biodiversity enhances bees and other insect pollinators in agroecosystems: a review." *Agronomy for Sustainable Development* 33, 257–74.

Oldroyd B. (2007) "What's killing American honey bees?" *PLoS Biology* 5, 1195–9.

Oldroyd B. P., Nanork P. (2009) "Conservation of Asian honey bees." *Apidologie* 40, 296–312.

Ollerton J., Winfree R., Tarrant S. (2011) "How many flowering plants are pollinated by animals?" *Oikos* 120, 321–6.

Osborne J. L. (2012) "Bumblebees and pesticides." *Nature* 491, 43–5.

Palmer K., Oldroyd B. (2000) "Evolution of multiple mating in the genus *Apis.*" *Apidologie* 31, 235–48.

Pettis J. S., Lichtenberg E. M., Andree M., et al. (2013) "Crop pollination exposes honey bees to pesticides which alters their susceptibility to the gut pathogen Nosema ceranae." *PLoS ONE* 8, e70182.

Potts S. G., Biesmeijer J. C., Kremen C., et al. (2010) "Global pollinator declines: trends, impacts and drivers." *Trends in Ecology and Evolution* 25, 345–53.

Ratnieks F. L. W., Carreck N. L. (2010) "Clarity on honey bee collapse?" *Science* 327, 152–3.

Renner M. A., Nieh J. C. (2008) "Bumble bee olfactory information flow and contact-based foraging activation." *Insectes Sociaux* 55, 417–24.

Roberts S., Potts S., Biesmeijer K., et al (2011) "Assessing continental scale risks for generalist and specialist pollinating bee species under climate change." *BioRisk* 6, 1–18.

Roubik D. W. (2006) "Stingless bee nesting biology." *Apidologie* 37, 124–43.

Sammataro D., Gerson U., Needham G. (2000) "Parasitic mites of honey bees: life history, implications, and impact." *Annual Review of Entomology* 45, 519–48.

Simone M., Evans J., Spivak M. (2009) "Resin collection and social immunity in honey bees." *Evolution* 63, 3016–22.

Starks P. T., Blackie C., Seeley T. (2000) "Fever in honeybee colonies." *Naturwissenschaften* 87, 229–31.

Steinhauer N. A. et al. (2014) "A national survey of managed honey bee 2012–2013 annual colony losses in the USA: results from the Bee Informed Partnership." *Journal of Apicultural Research* 53 (1).

Stow A., Briscoe D., Gillings M., et al. (2007) "Antimicrobial defences increase with sociality in bees." *Biology Letters* 3, 422–4.

Vanbergen A. J. and the Insect Pollinators Initiative (2013) "Threats to an ecosystem service: pressures on pollinators." *Frontiers in Ecology and the Environment* 11, 251–9.

van Engelsdorp D., Meixner M. D. (2010) "A historical review of managed honey bee populations in Europe and the United States and the factors that may affect them." *Journal of Invertebrate Pathology* 103, 580–95.

Vasquez A., Forsgren E., Fries I., et al. (2012) "Symbionts as major modulators of insect health: lactic acid bacteria and honeybees." *PLoS ONE* 7, e33188.

Weinstock G., Robinson G., Gibbs R., et al. (2006) "Insights into social insects from the genome of the honeybee *Apis mellifera*." *Nature* 443, 931–49.

Whitehorn P. R., O'Connor S., Wackers F. L., Goulson D. (2012) "Neonicotinoid pesticide reduces bumblebee colony growth and queen production." *Science* 336, 351–2.

Williams P. H., Osborne J. L. (2009) "Bumblebee vulnerability and conservation worldwide." *Apidologie* 40, 367–87.

Williamson S. M., Wright G. A. (2013) "Exposure to multiple cholinergic pesticides impairs olfactory learning and memory in honey bees." *Journal of Experimental Biology* 216, 1799–807.

Wilson-Rich N., Spivak M., Fefferman N., Starks P. (2009) "Genetic, individual, and group facilitation of disease resistance in insect societies." *Annual Review of Entomology* 54, 405–23.

WEBSITES

Atlas Hymenoptera (University of Mons). zoologie.umh.ac.be/hymenoptera

Australian Native Bee Research Centre. www.aussiebee.com.au

BeeBase. www.beebase.org

Bee Informed Partnership. www.beeinformed.org

Bees Wasps & Ants Recording Society. www.bwars.com

BugGuide. www.bugguide.net

Classroom Observation Hives. www.classroomhives.org

Discover Life. www.discoverlife.org

National Honey Board (USA). www.honey.com

Natural History Museum, London. www.nhm.ac.uk

Xerces Society for Invertebrate Conservation. www.xerces.org

Index

diseases 27, 42–43, 126, 134–37, 188–89
dominance 56, 65
domino cuckoo bee 152
drones 9, 28, 33, 44, 85
Dufour's gland 30, 56, 153, 160
dwarf honey bee 183

E

eastern cucurbit bee 149
ecological constraints hypothesis 57
endocrine system 38–39
Eucerini *see* long-horned bees
Eufriesea auripes 151
Euglossine bees *see* orchid bees
European foulbrood (EFB) 42, 134–35, 136, 137
eusociality 14, 53, 54–55, 56, 57, 82
evolution 6, 14, 20, 23, 45, 54–55
    see also natural selection
eyes 10, 29, 32, 33, 75

F

fear of bees 206–7
fertilizers 195, 197
flight 31, 81
food production 106–7
foraging 9, 14, 66, 68–69, 75, 79
foulbrood 43, 211
    AFB 42, 126, 134, 136, 137, 211
    EFB 42, 134–35, 136, 137
Francis de Sales, St. 101
fungi 42, 135, 137, 202, 203
fungicides 127, 130, 201

G

garden bumble bee 166
genetically modified crops 198
genetics 34–35
genomics 10, 20, 36–37, 40
giant honey bee 80, 182
Gobnait, St. 103
Gregory, St. 103
guard bees 39, 54, 59, 79, 80, 131, 179

H

habitat loss 27, 192–93, 212
hairy-footed flower bee 52, 68, 73, 87, 147
Halictidae *see* sweat bees
Hamilton, W. D. 45, 57
hand pollination 27, 107
haplodiploidy 34, 35, 45
harvesting 124–25
hearing 33
herbicides 127, 130, 197, 198, 201
Hippocrates 92
hive beetles 43, 132–33, 136, 137
hives 10, 16, 81, 114–19, 130–31
Hölldobler, Burt 59
honey 6, 9, 10, 16, 70–71, 92, 97, 98, 100, 101, 124–25, 127, 173
honey bees 16, 19, 20–21, 23, 26, 34, 37, 43, 46, 47, 49, 53, 65, 74, 80, 85, 180–85, 188, 205
    subspecies 131
hormones 38–39
Huber, François 116
humoral immunity 41
hyperpolyandry 88–89

I

immunology 40–41
informatics 36–37
insecticides *see* pesticides
integrated management 130–31
irapuá bee 177
iratim bee *see Lestrimelitta limao*
Islam 100
Isle of Wight disease 188, 209

J

jetaí 179
Judaism 98, 101
juvenile hormone (JH) 38–39

K

Kharlamii, St. 103
kleptoparisitism 19
Koschevnikov's bee 184

L

Langstroth hives 116–17, 118
larvae 9, 10, 20, 42, 47, 82, 83
leafcutter bees 19, 107, 113, 141, 159
legs 28
lemon straw bee 178
Lestrimelitta limao 80, 178, 179
life cycle 46–47
lifespan 9, 21, 27, 38, 39
long-horned bees 20, 22, 149
loosestrife bees 22

Author Biographies

CHAPTERS 1 TO 3, 5

**by Dr. Noah Wilson-Rich
& Kelly Allin**

Noah Wilson-Rich is the founder and Chief Scientific Officer of The Best Bees Company, a beekeeping service and research organization based in Boston, Massachusetts. He is a TED speaker and an expert in urban beekeeping, and the main focus of his research is on improving honey bee health. Noah has a PhD from Tufts University, awarded for research on genetic, individual, and group facilitation of disease resistance in the honey bee and two species of paper wasp.

Kelly Allin was the first Laboratory Manager at The Best Bees Company's Urban Beekeeping Laboratory & Bee Sanctuary in Boston. Kelly's contributions to this book overlapped with her studies while pursuing a biology degree at Northeastern University. Kelly is dedicated to all things related to sustainability and urban agriculture.

CHAPTERS 6 & 7

**by Norman Carreck
& Dr. Andrea Quigley**

Norman Carreck has been keeping bees for over thirty years, and has been a bee research scientist for more than twenty. He has given lectures about bees on all continents where bees are kept, has written many scientific papers, book chapters, and popular articles, has edited several books, and has made regular media appearances. He is Science Director of the International Bee Research Association and senior editor of the Journal of Apicultural Research.

Andrea Quigley has a BSc in applied biology from Hertfordshire University and a PhD in agricultural botany from the University of Wales, Aberystwyth, UK, where she was based at the Welsh Plant Breeding Station. She has kept bees for over ten years and holds the BBKA Basic Certificate in beekeeping competence. A freelance writer, she writes regularly for beekeeping journals on plants for bees, and on all aspects of bees and beekeeping.

Chapter 4 was written by all the authors.

Acknowledgments 🐝

Noah Wilson-Rich would like to thank Lea Campolo for her research into bees and spirituality and Jeff Murray (Classroom Hives) and Benadette Manning (Boston Public Schools) who provided research into and advice about observation hives. Noah would also like to thank Jacqueline Beaupre, Bryan Wilson-Rich, Mark Lewis, and Kristian Demary for their valuable research contributions.

Norman Carreck would like to thank Mark Greco (University of Bath) for supplying reference photos for stingless bees.

The authors would also like to thank David Price-Goodfellow and Hugh Brazier for their helpful and constructive editing and guidance.

Ivy Press would like to thank the following for permission to reproduce copyright material:

Alamy/FLPA: 72; Nick Michaluk: 89; Carver Mostardi: 7; Prisma Bildagentur AG: 128; WildPictures: 19R

Bigstock: 96, 97T, 97C, 106

Bridgeman Art Library/M.H. de Young Memorial Museum, San Francisco, CA, USA/Alinari: 100; Museum of Fine Arts, Boston, Massachusetts, USA/Henry Lillie Pierce Fund: 98; Stapleton Collection: 17

Corbis/Kristian Buus/In Pictures: 125B; Roberta Olenick/All Canada Photos: 84; Eric Tourneret/Visuals Unlimited: 86, 111T, 180

Maja Dumas: 125T

FLPA/Richard Becker: 82B, 113; Biosphoto: 191; Heidi & Hans-Juergen Koch/Minden Pictures: 39

Fotolia: 29, 145, 152, 185

Getty Images/BlackCatPhotos: 45; The British Library/Robana: 92; Thierry Charlier/AFP: 201; De Agostini: 23; Loomis Dean/Time Life Pictures: 199; Alfred Eisenstaedt/Time & Life Pictures: 200; Chris McLoughlin: 138; Robert Nickelsberg: 213B; Joe Raedle: 95; Chico Sanchez: 108; Gerard Sioen/Gamma-Rapho: 99; Universal Images Group: 16T; Visuals Unlimited: 11; Visuals Unlimited, Inc./John Abbott: 55; Visuals Unlimited, Inc./Eric Tourneret: 56

David B. Gleason: 122

Eliza Grinnell/Ben Finio: 94

ImagineChina/Li Junsheng:: 107

iStockphoto: 58, 59, 93, 124, 194

Rebecca Leaman: 77

Library of Congress: 196

Courtesy of the John D. & Catherine T. MacArthur Foundation: 126

© The Trustees of the Natural History Museum, London: 14; photographed by Harry Taylor: 142, 143, 144, 148, 149, 150, 151, 154, 155, 156, 157, 159, 160, 161, 162, 171, 174, 175, 176, 177, 178, 179, 181, 182, 183, 184

Nature Picture Library/Ingo Arndt: 12; Neil Bromhall: 5, 21; Simon Colmer: 54, 63; Laurent Geslin: 47, 123B; Chris Gomersall/2020VISION: 68; Dietmar Nill: 57T; Kim Taylor: 61, 173

NGA Images: 102

Masato Ono, Tamagawa University (Tokyo): 81B

Oxford University Images: 158

Jerry A. Payne: 15

Jerome G. Rozen: 73B

Sailko: 104T

Gilles San Martin: 18, 22, 133

Jordan Schwartz: 115B

Science Photo Library/Valerie Giles: 83T; Cordelia Molloy: 33B; Eric Tourneret/Visuals Unlimited: 26

Shutterstock: 9, 18, 19L, 24, 37, 40, 48, 49, 50, 52, 64, 70, 78, 79, 80, 97B, 110, 117L, 117R, 120, 127, 135, 140, 146, 147, 163, 165, 168, 169, 170, 172, 192, 193T, 195, 196, 198, 205B, 206, 207, 209, 213T

Arnstein Staverløkk/Norwegian Institute for Nature Research: 153, 166, 167

Shinya Suzuki: 123T

United States Department of Agriculture: 22, 27, 36, 43L, 43R, 43C, 119, 188, 189, 202, 203, 204, 210, 211

USGS Bee Inventory and Monitoring Lab/Sam Droege: 33, 69, 132B

Waugsberg: 53, 71

Paul Zborowski: 73T, 75, 85